Springer Theses

Recognizing Outstanding Ph.D. Research

Aims and Scope

The series "Springer Theses" brings together a selection of the very best Ph.D. theses from around the world and across the physical sciences. Nominated and endorsed by two recognized specialists, each published volume has been selected for its scientific excellence and the high impact of its contents for the pertinent field of research. For greater accessibility to non-specialists, the published versions include an extended introduction, as well as a foreword by the student's supervisor explaining the special relevance of the work for the field. As a whole, the series will provide a valuable resource both for newcomers to the research fields described, and for other scientists seeking detailed background information on special questions. Finally, it provides an accredited documentation of the valuable contributions made by today's younger generation of scientists.

Theses are accepted into the series by invited nomination only and must fulfill all of the following criteria

- They must be written in good English.
- The topic should fall within the confines of Chemistry, Physics, Earth Sciences, Engineering and related interdisciplinary fields such as Materials, Nanoscience, Chemical Engineering, Complex Systems and Biophysics.
- The work reported in the thesis must represent a significant scientific advance.
- If the thesis includes previously published material, permission to reproduce this must be gained from the respective copyright holder.
- They must have been examined and passed during the 12 months prior to nomination.
- Each thesis should include a foreword by the supervisor outlining the significance of its content.
- The theses should have a clearly defined structure including an introduction accessible to scientists not expert in that particular field.

More information about this series at http://www.springer.com/series/8790

Robert J.A. Francis-Jones

Active Multiplexing of Spectrally Engineered Heralded Single Photons in an Integrated Fibre Architecture

Doctoral Thesis accepted by
the University of Bath, UK

 Springer

Author
Dr. Robert J.A. Francis-Jones
Department of Physics
University of Bath
Bath
UK

Supervisor
Dr. Peter J. Mosley
Department of Physics
Centre for Photonics and Photonic Materials,
University of Bath
Bath
UK

ISSN 2190-5053   ISSN 2190-5061 (electronic)
Springer Theses
ISBN 978-3-319-64187-4   ISBN 978-3-319-64188-1 (eBook)
DOI 10.1007/978-3-319-64188-1

Library of Congress Control Number: 2017948194

© Springer International Publishing AG 2017
This work is subject to copyright. All rights are reserved by the Publisher, whether the whole or part of the material is concerned, specifically the rights of translation, reprinting, reuse of illustrations, recitation, broadcasting, reproduction on microfilms or in any other physical way, and transmission or information storage and retrieval, electronic adaptation, computer software, or by similar or dissimilar methodology now known or hereafter developed.
The use of general descriptive names, registered names, trademarks, service marks, etc. in this publication does not imply, even in the absence of a specific statement, that such names are exempt from the relevant protective laws and regulations and therefore free for general use.
The publisher, the authors and the editors are safe to assume that the advice and information in this book are believed to be true and accurate at the date of publication. Neither the publisher nor the authors or the editors give a warranty, express or implied, with respect to the material contained herein or for any errors or omissions that may have been made. The publisher remains neutral with regard to jurisdictional claims in published maps and institutional affiliations.

Printed on acid-free paper

This Springer imprint is published by Springer Nature
The registered company is Springer International Publishing AG
The registered company address is: Gewerbestrasse 11, 6330 Cham, Switzerland

*For Mum,
and
Tracey.*

Supervisor's Foreword

The recent explosion in the field of quantum-enhanced technology has been driven by the promise of step changes in the way in which information is processed. The laws of quantum mechanics enable sensing, communication and—in particular—computation to be carried out in qualitatively more powerful ways if we can harness ensembles of individual quantum objects. For example, provably secure communication is possible by transmitting a cryptographic key using single photons, and quantum computers could achieve huge speed-up in tasks such as factorisation. Whilst it seems unlikely at the moment that quantum processors will become as ubiquitous as the microprocessors that pervade everyday life in the early twenty-first century, nevertheless for high-end calculations including complex simulations they would be revolutionary.

Many different platforms are currently competing to lead the race towards 'quantum supremacy'—the point at which a quantum processor can carry out calculations more rapidly than its classical counterpart. Many of these platforms are highly technically complex, for example trapped ion processors requiring many laser systems and ultra-high vacuum, or superconducting devices at cryogenic temperatures. Photonics—in which single photons are used as the information carriers—has various practical advantages; in particular at ambient temperature and pressure, photons do not suffer decoherence from interaction with the environment. Advanced photonic quantum processors can in principle be constructed from single-photon sources, logic gates and photon detectors. However, due to the weakness of nonlinear interactions between photons, photonic quantum logic gates rely on Hong–Ou–Mandel interference between two photons, and the overall speed of a photonic quantum processor is critically dependent on photon loss.

This thesis is concerned with another difficulty of using photons for quantum information processing—that of generating single photons in the first place. Reliable single-photon sources are notoriously difficult to build, and a photonic quantum processor requires many of them to operate simultaneously. Importantly, each source must also produce photons that are identical and capable of high-visibility interference—that is to say in a pure quantum state. The work reported in this thesis demonstrates how the delivery probability of high-purity photons can be increased using active routing.

Dr. Francis-Jones designed and built a single-photon source based on photon-pair generation by spontaneous four-wave mixing. Four-wave mixing is a process that can take place as a high-intensity pulse of light propagates along an optical fibre, generating frequency sidebands above and below the pump. Fabricating bespoke photonic crystal fibre (PCF), in which light guidance is controlled by a matrix of air holes running along the length of the fibre, enabled Dr. Francis-Jones to use the technique of group velocity matching between the pump laser pulses and the photon pairs generated to create heralded photons in high-purity quantum states. However, imperfections in the PCF due to the fabrication process required him to characterise variations in the PCF along its length to a much higher degree of accuracy than had been done previously and carry out a numerical reconstruction of the PCF profile. The PCF could then be built into a fully spliced fibre architecture, along with additional fibre components to isolate the generated photon pairs from the residual pump laser, including bandgap-guiding PCF fabricated to transmit only the frequency bands of interest.

The latter chapters of the thesis show how the output from two identical sources was combined to increase the reliability of the resulting single-photon source. Photon-pair generation is spontaneous and typically occurs only from a small fraction of the pump laser pulses. However, because the photons are paired, by pumping two or more sources simultaneously and splitting up the pairs created, when one photon is detected the other can be routed to a single output using optical delay and a fast switch. This technique is known as multiplexing and is the primary candidate for building high-reliability single-photon sources; under perfect conditions, the output of seventeen sources provides over 99% delivery probability. The construction of an all-fibre source of high-purity photons that required little optical alignment brought the possibility of actively linking two—or many more—sources to increase the overall delivery rate of single photons. Dr. Francis-Jones successfully multiplexed two identical sources through a fast, low-loss, fibre-coupled optical switch and demonstrated a clear improvement in performance over that of a single source.

The resulting source is the first to combine high-purity single-photon generation, a fully integrated architecture and an increase in the single-photon delivery probability through multiplexing. It is another step along the path to useful integrated quantum photonic devices, as well as a significant scientific achievement in its own right.

Bath, UK
May 2017

Dr. Peter J. Mosley

Abstract

In recent years, there has been rapid development in processing of quantum information using quantum states of light. The focus is now turning towards developing real-world implementations of technologies such as all-optical quantum computing and cryptography. The ability to consistently create and control the required single-photon states of light is crucial for successful operation. Therefore, high-performance single-photon sources are very much in demand.

The most common approach of generating the required nonclassical states of light is through spontaneous photon pair generation in a nonlinear medium. One photon in the pair is detected to 'herald' the presence of the remaining single photon. For many applications, the photons are required to be in pure indistinguishable states. However, photon pairs generated in this manner typically suffer from spectral correlations, which can lead to the production of mixed, distinguishable states. Additionally, these sources are probabilistic in nature, which fundamentally limits the number of photons that can be delivered simultaneously by independent sources and hence the scalability of these future technologies.

One route to deterministic operation is by actively multiplexing several independent sources together to increase the probability of delivering a single photon from the system. This thesis presents the development and analysis of a multiplexing scheme of heralded single photons in high-purity indistinguishable states within an integrated optical fibre system. The spectral correlations present between the two photons in the pair were minimised by spectrally engineering each photonic crystal fibre source. A novel, in-fibre, broadband filtering scheme was implemented using photonic bandgap fibres. In total, two sources were multiplexed using a fast optical switch, yielding an 86% increase in the heralded count rate from the system.

Publication List

1. Robert J.A. Francis-Jones, Rowan A. Hoggarth and Peter J. Mosley, *All-fibre multiplexed source of high-purity single photons*, Optica, **3**(11), pp. 1270–1273 (2016).

2. Robert J.A. Francis-Jones and Peter J. Mosley, *Characterisation of longitudinal variation in photonic crystal fibre*, Optics Express, **24**(22), pp. 24863–24845 (2016).

3. Robert J.A. Francis-Jones and Peter J. Mosley, *An all-fibre low-noise source of pure heralded single photons*, In Preparation (2016).

4. Robert J.A. Francis-Jones, Rowan A. Hoggarth and Peter J. Mosley, *Spatial multiplexing of high-purity heralded single photons in an integrated fibre architecture*, in *CLEO: 2016*, p. FTh1C.3, Optical Society of America (2016).

5. Robert J.A. Francis-Jones and Peter J. Mosley, *Experimental characterisation of longitudinal uniformity in photonic crystal fibre*, in *CLEO: 2016*, p. STu4P.4, Optical Society of America (2016).

6. Robert J.A. Francis-Jones and Peter J. Mosley, *Temporal loop multiplexing: A resource efficient scheme for multiplexed photon-pair sources*, arXiv e-prints arXiv:1503.06178 (2015).

7. Peter J. Mosley, Itandehui Gris-Sánchez, James M. Stone, Robert J. A. Francis-Jones, Douglas J. Ashton, and Tim A. Birks, *Characterizing the variation of propagation constants in multicore fibre*, Optics Express, **22**, pp. 25689–25699 (2014).

8. Oliver J. Morris, Robert J.A. Francis-Jones, Keith G. Wilcox, Anne C. Tropper and Peter J. Mosley, *Photon-pair generation in photonic crystal fibre with a 1.5 GHz modelocked VECSEL*, Optics Communications, **327**, pp. 39–44, *Special issue on Nonlinear Quantum Photonics* (2014).

9. Robert J.A. Francis-Jones and Peter J. Mosley, *Exploring the limits of multiplexed photon-pair sources for the preparation of pure single-photon states*, arXiv e-prints arXiv:1409.1394 (2014).

Acknowledgements

I would first like to express my gratitude to my supervisor Dr. Peter Mosley, for giving me the opportunity to work with him on an exciting and interesting project, and for first introducing me to the world of quantum optics in the final year of my undergraduate degree. Without his support and guidance over the last three years, this thesis and the work that it contains would not have been possible. It has been a tremendously enjoyable learning experience (for both of us I hope!), and I think I can safely say that I now know more about FPGAs than I ever thought or hoped I would! I am incredibly grateful for the opportunity and eagerly look forward to continue working with you on the upcoming NQIT project.

Secondly, I would like to thank all of the staff and students, past and present, of the Centre for Photonics and Photonic Materials for providing an enjoyable, interesting and supportive learning environment. I would particularly like to thank James, for casting his eye over my fibre designs, passing down his fabrication knowledge and teaching me the dark art of the dispersion rig! I would also especially like to thank Rowan for his help with running the fibre tower during the seemingly unending saga of PCF fabrication, it was a tough one but we got there in the end! I would also like to thank Stephanos for producing the tapers used in the fibre filtering scheme, and Simon for putting up with all of my unusual I.T. requests and breaking of my computer whilst trying to get the FPGA and PC on speaking terms with each other.

Thirdly, I would like to thank all of the friends I have made over the last three years, the list is long but that is only because you have all been so welcoming and supportive over the course of my Ph.D. studies. The climbing tie-dye crew of James, Ian, Miranda, Hazel and Emily, and also those within the Department of Physics: Duncan, Rose, Adrian, Rowan, Stephanos, Hasti, Kerri and Kat, thank you all for providing a welcome distraction from the broken lasers and malfunctioning electronics over the years!

Finally, my everlasting gratitude goes to my Mum, Alexia, for always providing support and encouragement to me. I hope that this goes some way to repaying you for everything you have done for me and proving that it was all worth it in the end. It is to you, and also Tracey, that this thesis is dedicated.

Contents

1 Introduction.. 1
   1.1 What Is a Photon?....................................... 3
      1.1.1 Indistinguishability and Hong-Ou-Mandel Interference.... 4
   1.2 Methods of Single Photon Generation...................... 7
      1.2.1 Single Emitter Sources............................ 7
      1.2.2 Heralded Single Photon Sources.................... 9
   1.3 Multiplexing: A Route to Deterministic Operation........... 10
   1.4 Thesis Outline.. 15
   References.. 16

2 Photon Pair Generation via Four-Wave Mixing in Photonic
   Crystal Fibres... 21
   2.1 Overview.. 21
   2.2 Photon Pair Generation in Photonic Crystal Fibres.......... 23
   2.3 Spectral Engineering of the Two-Photon State.............. 26
   2.4 Reduced State Spectral Purity: Schmidt Decomposition...... 31
   2.5 Photon Number Statistics................................ 32
      2.5.1 Raman Scattering................................. 32
      2.5.2 Degree of Heralded Second Order Coherence........... 33
      2.5.3 Degree of Marginal Second Order Coherence........... 34
   References.. 35

3 Numerical Modelling of Multiplexed Photon Pair Sources......... 39
   3.1 Overview.. 39
   3.2 The Building Blocks.................................... 40
      3.2.1 Pair Generation................................... 40
      3.2.2 Detection.. 41
      3.2.3 Metrics and Measurements......................... 43
   3.3 Spatial Multiplexing.................................... 44
      3.3.1 Simulation Results and Discussion................... 47

	3.4	Temporal Loop Multiplexing	55
		3.4.1 Simulation Results and Discussion	58
	3.5	Conclusion	65
	References	65
4	**Design, Fabrication, and Characterisation of PCFs for Photon Pair Generation**	67
	4.1	Overview	67
	4.2	Photonic Crystal Fibre Design and Fabrication	68
		4.2.1 Simulation and Design	69
		4.2.2 PCF Fabrication and Characterisation	72
	4.3	Measuring the Joint Spectral Intensity Distribution	76
		4.3.1 Stimulated Emission Tomography of FWM	77
	4.4	The Effect of PCF Inhomogeneity on the Reduced State Purity	82
		4.4.1 Numerical Reconstruction of Inhomogeneous PCFs	83
		4.4.2 The Final Fibres	92
	4.5	Conclusions	93
	References	94
5	**Construction of an Integrated Fibre Source of Heralded Single Photons**	95
	5.1	Overview	95
	5.2	Photonic Bandgap Fibre Filter Design and Fabrication	96
	5.3	Component Assembly and Splicing	100
		5.3.1 PCF-SMF Splicing	101
		5.3.2 PBGF-SMF Splicing	101
		5.3.3 Assembly of the Photon Pair Source	105
	5.4	Detection and Coincidence Counting Electronics	108
	5.5	Optical Switch Integration	109
	References	113
6	**Characterisation of a Multiplexed Photon Pair Source**	115
	6.1	Overview	115
	6.2	Characterisation of the Coincidence Count Rates	115
		6.2.1 Coincidence Counting and Coincidence-to-Accidentals	117
		6.2.2 Experimental Results: Coincidences and CAR	118
		6.2.3 Experimental Results: Source 2 Noise Gating	124
	6.3	Characterisation of the Second Order Coherence	126
		6.3.1 Experimental Results: Marginal Second Order Coherence	127
		6.3.2 Experimental Results: Heralded Second Order Coherence	133
	6.4	Summary of Source Performance and Potential Improvements	135
	6.5	Conclusions	139
	References	139

7	**Conclusion**		141
	7.1	Summary	141
	7.2	Future Outlook	144
		7.2.1 Photon Pair Generation in Fibres	144
		7.2.2 Multiplexed Photon Pair Sources	145
	References		146

Acronyms

AlGaAs	Aluminium Gallium Arsenide
APD	Avalanche Photo-Diode
ARROW	Anti-Resonant Reflecting Optical Waveguide
BBO	Beta-Barium Borate
BFA	Bare Fibre Adapter
BiBO	Bismuth Triborate
BNC	Bayonet Neil–Concelman Connector
CAR	Coincidence-to-Accidentals Ratio
CW	Continuous Wave
DFG	Difference Frequency Generation
DM	Dichroic Mirror
DOS	Density of States
EO	Electro-Optic
EOM	Electro-Optic Modulator
FBG	Fibre Bragg Grating
FC	Ferrule Connector
FPGA	Field Programmable Gate Array
FWM	Four-Wave Mixing
GPIB	General Purpose Interface Bus
GVM	Group Velocity Matched
HBT	Hanbury Brown–Twiss
Hi1060	Single-Mode Fibre at 1060 nm
HOM	Hong–Ou–Mandel Interference
HWP	Half-Wave Plate
InGaAs	Indium Gallium Arsenide
JSA	Joint Spectral Amplitude
JSI	Joint Spectral Intensity
KLM	Knill, Laflamme and Milburn
LMA	Large Mode Area
LOQC	Linear Optics Quantum Computing

MDF	Medium Density Fibreboard
MFD	Mode Field Diameter
NIM	Nuclear Instrumentation Modules
OSA	Optical Spectrum Analyser
PBGF	Photonic Bandgap Fibre
PBS	Polarising Beamsplitter
PC	Personal Computer
PCF	Photonic Crystal Fibre
PDC	Parametric Downconversion
PhCW	Photonic Crystal Waveguide
PMC	Phase-matching Contour
PMF	Phase-matching Function
PNR	Photon Number Resolving
POVM	Positive Operator Value Measure
PPLN	Periodically Poled Lithium Niobate
PPLNW	Periodically Poled Lithium Niobate Waveguide
RMS	Root Mean Square
SEM	Scanning Electron Micrograph
SET	Stimulated Emission Tomography
SM800	Single-Mode Fibre at 800 nm
SMF	Single-Mode Fibre
SMF28	Single-Mode Fibre at 1550 nm
SNR	Signal-to-Noise
SPM	Self-Phase Modulation
USB	Universal Serial Bus
VECSEL	Vertical External Cavity Emitting Laser
WDM	Wavelength Division Multiplexer
ZDW	Zero Dispersion Wavelength

Mathematical Variables

A_{eff}	Effective area of the waveguide mode
F	Fidelity of the heralded single-photon state
H_I	FWM interaction Hamiltonian
K	Schmidt number or cooperativity parameter
L	Fibre length
M	Number of modes in a pseudo-photon number resolving detector
N_m	Number of spectral modes present
N_p	Number of independent heralded single photons
N_s	Number of switches in 'log-tree' scheme
P	Reduced state purity
$P(n)$	Probability per pulse of delivering n photons
P_p	Pump field peak power
R_p	Pump laser repetition rate
$\Delta\beta$	Propagation constant phase mismatch
$\Delta\omega_{FWHM}$	Full width half maximum pump bandwidth
Δk	Phase mismatch
Δn	Standard deviation of photon number
Δp	Difference between $p(\text{success})$ for the analytical model of temporal multiplexing and a full numerical calculation using the density matrix
Λ	Pitch of PCF cladding
Ω	Phonon frequency
$\hat{a}_j$	Annihilation operator for the j'th field
$\bar{I}$	Time-averaged intensity
$\bar{\Lambda}$	Mean value of the PCF pitch
$\bar{n}$	Mean photon number
β	Propagation constant of the fundamental waveguide mode
β_1	Inverse group velocity
β_2	Group velocity dispersion
E	Electric field vector

$\mathbf{P}$	Material polarisation
$\chi^{(i)}$	i'th order dielectric susceptibility
$\hat{a}_j^\dagger$	Creation operator for the j'th field
ϵ_0	Permittivity of vacuum
η	FWM efficiency parameter
η_L	Storage loop efficiency
η_τ	Delay line efficiency
η_d	Detector efficiency
η_s	Switch efficiency
γ	Optical nonlinearity
$\hat{E}_j^{(+)}$	Positive frequency component of the quantised electric field of the j'th field
$\hat{E}_j^{(-)}$	Negative frequency component of the quantised electric field of the j'th field
$\hat{\Pi}(n)$	POVM for a detection outcome of 'n'
$\hat{\gamma}_{Multi}^{(2)}(0)$	Experimental heralded second-order coherence function for the multiplexed device
$\hat{\gamma}_{S_j}^{(2)}(0)$	Experimental heralded second-order coherence function for source j
$\hat{\gamma}_{marg.}^{(2)}(0)$	Experimental marginal second-order coherence function
$\hat{\rho}$	Two-photon state density matrix
$\hat{\rho}_i$	Reduced density matrix of the idler photon
$\hat{\rho}_s$	Reduced density matrix of the signal photon
$\hat{g}_H^{(2)}(0)$	Degree of heralded second-order coherence function
$\hat{n}$	Photon number occupation operator
$\hbar$	Planck's constant
$f(\omega_s, \omega_i)$	Joint spectral amplitude
$|f(\omega_s, \omega_i)|^2$	Joint spectral intensity
$|0\rangle$	Vacuum fock state
$|\Psi\rangle$	Two-photon state vector
$|\theta(\omega_s)\rangle$	Schmidt mode basis of signal arm
$|\zeta(\omega_i)\rangle$	Schmidt mode basis of idler arm
$|n\rangle$	Fock state containing exactly n photons
λ	Free space wavelength
λ_j	Schmidt mode weighting for the j'th pair of modes
$\mathcal{F}$	Minimisation function
$\mathcal{H}$	Two-photon state Hilbert Space
$\mathcal{H}_i$	Idler photon Hilbert Space
$\mathcal{H}_s$	Signal-photon Hilbert Space
$\mathcal{R}$	Reflection probability amplitude coefficient
$\mathcal{T}$	Transmission probability amplitude coefficient
v_j	Group velocity of j'th field
ω	Angular frequency
ω_{j0}	Central frequency of the j'th field

Mathematical Variables

ϕ_j	Detuning from perfectly phase-matched frequencies
$\phi(\omega_s, \omega_i)$	Phase-matching function
$\alpha(\omega_s + \omega_i)$	Pump envelope function
σ_p	$1/e$ pump bandwidth
σ_p^{RMS}	Root-mean-square pump bandwidth
τ	Time delay
τ_j	Group velocity mismatch between the pump and field j
θ_{pmf}	Phase-matching function orientation angle
$\tilde{\alpha}(t)$	Temporal profile of the pump pulse
$\vec{p}_H$	Heralding probability vector
$\vec{r}_T$	Transverse position coordinate
$\vec{r}$	Position coordinate
a_n	Probability amplitude coefficient for the generation of n photon pairs
c	Speed of light in vacuum
d	Hole diameter in PCF cladding
$f(\omega_j)$	Spectral distribution function for the j'th field
$g^{(2)}(0)$	Degree of second-order coherence at zero time delay
$g_{cl}^{(2)}(0)$	Classical degree of second-order coherence at zero time delay
$g_m^{(2)}(0)$	Marginal degree of second-order coherence at zero time delay
$g_{qu}^{(2)}(0)$	Quantum degree of second-order coherence at zero time delay
k_j	Wavevector of the j'th field
n	Photon number
n_2	Kerr-coefficient of fused silica
n_{eff}	Effective refractive index of the waveguide mode
$p(N_p)$	Probability of delivering N_p photons
$p(\text{herald})$	Probability of a successful heralding event
$p(\text{noise})$	Probability of delivering more than one photon from a successful heralding signal
$p(\text{success})$	Probability of successfully delivering a single Photon
p_{acc}	Probability per pulse of an accidental coincidence detection event between the signal and idler arms
$p_{det}(n\|N)$	Conditional probability of the detector recording 'n' from 'N' photons at the input
p_{idl}	Probability per pulse of a detection event in the idler arm
p_{sig}	Probability per pulse of a detection event in the signal arm
p_{th}	Thermal photon number probability distribution
t	Time coordinate
t_{wait}	Waiting time
z	Longitudinal position coordinate

Chapter 1
Introduction

Overview and Motivation

Historically single photons and sources of single photons have been instrumental in helping to explore, understand and develop new theories and devices across a range of topics in quantum mechanics and quantum information science. In 1974 Clauser et al. demonstrated the first source of single photon emission from a cascade transition in calcium atoms [1]. Whilst it was commonly accepted at the time that a single photon was indivisible, and that if one were to send a single photon to a half-silvered mirror one would never measure a correlated detection signal between detectors placed in the transmitted and reflected arms, the experiments of Clauser were the first conclusive demonstration of such an effect.

This photon antibunching was also subsequently observed in the fluorescence from a sodium atomic beam by Kimble et al. [2] and Walls [3]. The atomic cascade sources provided access to states of light that had previously been inaccessible. Using these sources, Clauser and Shimony [4] and Aspect et al. [5] demonstrated the first violation of Bell's inequalities a test of local realism which was shown to be inconsistent with the developing theories of quantum mechanics. Although these sources were instrumental in several truly ground breaking experiments in quantum mechanics, the emission rates were low and the complexity of the apparatus high.

The advent of the laser by Maiman [6] in 1960 revolutionised the field of quantum optics. With the high-power and coherent beams from a laser it became possible to explore the non-linear optical properties of materials and in doing so the first sources of correlated photon pairs were produced by Burnham and Weinberg [7]. These photon pairs generated by parametric downconversion in non-linear crystals were produced at much higher rates compared to the previous atomic cascade sources. A single photon state could be realised from the pair by first separating the two photons and detecting one of them, such as in the experiments of Hong and Mandel [8]. The number correlation between the photons in the pair meant that whenever this detector fired the presence of the remaining photon was "heralded".

These new and improved sources were put to great effect and further advances in the field of quantum optics were made. Notably by Hong, Ou and Mandel who in 1987 demonstrated the now ubiquitous two-photon interference effect at a beam splitter that was subsequently named after them [9]. This effect and measurement technique has become one of the gold standards by which the quality of the output from a single photon source is judged. Additionally, it is key to operation of many optical implementations of quantum logic gates as it provides a means by which two photons can be easily entangled.

Parametric downconversion sources of single photons have become the workhorses of the quantum optics community, as they are relatively simple to construct and are capable of producing single photons at high rates, the only downside is that the generation mechanism is not deterministic. This makes it difficult to generate multiple photons from multiple independent sources simultaneously, although there are potential solutions on the horizon such as active multiplexing [10]. Nevertheless, Yao et al. have produced an eight photon entangled state by using four parametric downconversion sources pump simultaneously by a common laser system, albeit with a low eight-photon count rate [11].

Because of the relative ease with which photons can be generated (with engineered properties) in these schemes, quantum photonics is a promising candidate for the development of new technologies that make use of quantum mechanical effects to achieve some performance enhancement, such as in quantum computing, quantum cryptography and quantum metrology [12]. In the optical implementation of these schemes, photons are used as the information carriers required to encode and process quantum information [13]. Photons are particularly appealing as information carriers, as they interact only very weakly with the surrounding environment making them relatively robust to decoherence, and therefore any information encoded in the quantum state of the photon is not corrupted or lost.

Optical implementations of quantum computing such as linear optics quantum computing (LOQC) [14] or cluster state quantum computing [15] make use of photons in precisely controlled states [16]. Information may be encoded in the polarisation state of the photon, but also in superposition of paths taken, with conversion between the two made possible using only simple optics. Even if the photon is not used for information processing, a photonic "flying" qubit is unparalleled in its ability to transfer information over long distances. This is especially true when the wavelength of the photon lies in one of the low loss telecommunications bands of the existing optical fibre network that supports the current information technology infrastructure. This opens the possibility for using photonic qubits for encryption keys in quantum cryptography, allowing provably secure communication between different locations [17, 18], or as optical interconnects between quantum "processors" in a distributed quantum processor [19].

The uses of photons are not just limited to emerging technologies, but also allow fundamental tests into the nature of quantum mechanics such as through the violation of Bell inequalities [20–24]. In order to access these groundbreaking techniques an understanding of the method of generating non-classical states of light is required, such as entangled-pairs and indistinguishable single photons. The most widespread sources in use today are those based around nonlinear photon pair generation, but these sources are probabilistic. A photon-pair may only be generated within the medium with a given probability, determined in part by the strengths of the pump laser and the optical nonlinearity of the material. This is typically kept small in an effort to reduce the level of noise from other competing nonlinear processes. Thus, there is no prior knowledge of when a generation event will occur and the majority of pump pulses result in no pair generation at all. To fully realise the potential of these quantum technologies deterministic sources of well-controlled single photons will be required. This thesis explores how this may be achieved using active multiplexing.

1.1 What Is a Photon?

The term photon is used to describe a single excitation of the electromagnetic field [25]. A single photon state can be written in terms of the creation and annihilation operators of the j'th mode of the field [26]:

$$|1\rangle = \hat{a}_j^\dagger |0\rangle, \quad (1.1)$$
$$|0\rangle = \hat{a}_j |1\rangle, \quad (1.2)$$

where $|1\rangle$ is the state vector describing a mode containing only one photon, and $|0\rangle$ represents the mode containing no photons, also called the vacuum state. This state is not occupied, but the expectation value of the energy in the mode is non-zero due to vacuum fluctuation. This gives rise to the "zero-point" energy, so no mode of the quantised field is every truly empty but is occupied by $\hbar\omega/2$ [27]. It is this zero-point energy that allows the spontaneous generation of photon pairs in a parametric interaction within a medium, as will be seen later in this thesis.

Whilst convenient to write down, this description of a single photon state is incomplete. It does not describe any of the properties that are required of a single photon such as its frequency. This can be incorporated, by including the frequency dependence (or labelling) of the creation and annihilation operators for the spectral mode which the photon occupies [26]:

$$|\omega_j\rangle = \hat{a}_j^\dagger(\omega_j)|0\rangle. \quad (1.3)$$

This describes a single photon in the j'th spectral mode with a well-defined single frequency ω_j.

However, this leads to a second new problem. Photons do not have a single well-defined frequency. A single frequency dictates that the photon has an infinite temporal duration, this is at odds with the traditional particle-like view of a photon as an individual wavepacket [26, 27]. The photon must therefore have some degree of spectral structure. This is incorporated by performing a weighted sum over a number of spectral modes of the quantised field, using the spectral distribution function $f(\omega_j)$ [28, 29],

$$|\psi(\omega_j)\rangle = \int d\omega_j f(\omega_j) \hat{a}_j^\dagger(\omega_j) |0\rangle. \tag{1.4}$$

But this still only defines the spectral (and temporal) degrees of freedom. This discussion can be extended to also include the momentum and polarisation distribution [27]. It is the description of the photon in Eq. 1.4, that will be used throughout the remainder of this thesis.

The formalism developed above shows that despite its quantised particle-like nature, a photon possesses neither a localised position in time or space, but is rather distributed over, a potentially broad, range of both [27]. However, the photon is indivisible and so on detection will collapse into a single field excitation at one location.

The properties of the ideal source of single photons may vary depending on the final application of these states of light. In this thesis, single photon states that have potential uses in quantum information processing tasks are the primary focus. For this role, the single photons must be generated in a indistinguishable, pure state. This places quite stringent demands on the source itself.

Firstly, photons produced by independent sources, and indeed two or more photons generated successively by the same source, must be completely indistinguishable. To do this, one must be able to exercise absolute control over the degrees of freedom available to the photon at the point of generation, namely spectral and temporal distribution, spatial profile, momentum and polarisation. Secondly, the photons must be emitted into a pure quantum state, rather than an incoherent mixture of different modes. Finally, the photons should be produced on demand. When a photon is requested from the source, one and only one should be delivered from the output with unit probability, i.e. it should be deterministic. In reality there are a number of hurdles to overcome to achieve the three above criteria from a single source.

1.1.1 Indistinguishability and Hong-Ou-Mandel Interference

The lack of interaction with the environment that makes photons suitable for information transfer over large distances, is a limiting factor in the development of an all-optical quantum computer [13]. Interactions between the photonic qubits are too weak at the single photon level [13]. In 2001, Knill, Laflamme and Milburn (KLM) [14] showed, in what became known as the KLM implmentation of linear optics quantum computing, that a scalable photonic quantum computer is possible using only single-photon sources, linear optical circuits and single photon detectors. The KLM scheme is dependent on the Hong-Ou-Mandel (HOM) effect of quantum interference between two single photons at a beamsplitter.

1.1 What Is a Photon?

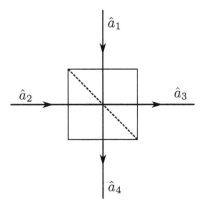

Fig. 1.1 Illustration of the input ($\hat{a}_1$, $\hat{a}_2$) and output ($\hat{a}_3$, $\hat{a}_4$) modes of a beamsplitter

The beamsplitter is a partially reflecting mirror, with two input ports and two output ports, illustrated schematically in Fig. 1.1. For a balanced or 50:50 beamsplitter, a photon incident into one of the input ports has a probability of 0.5 of exiting in either output port. In classical optics, the light impinging on the beamsplitter is redirected into the two output modes with the power split evenly between them. This effect underlays the second order coherence measurements in Chap. 6.

In general, the beamsplitter can be described by a reflection coefficient $\mathcal{R}^2$ and transmission coefficient $\mathcal{T}^2$. The beamsplitter operation acting on the input modes 1 and 2, transforming them to modes 3 and 4 is given by [26]

$$\hat{a}_3 = \mathcal{R}\hat{a}_1 + \mathcal{T}\hat{a}_2, \tag{1.5}$$

$$\hat{a}_4 = \mathcal{T}\hat{a}_1 - \mathcal{R}\hat{a}_2, \tag{1.6}$$

where $\hat{a}_j$ is the annihilation operator for the j'th mode of the beamsplitter [26]. The minus sign in Eq. 1.6 arises due to energy conservation ($\mathcal{R}^2 + \mathcal{T}^2 = 1$) and the π-phase shift acquired on reflection. From this the creation operators for the input modes 1 and 2 are:

$$\hat{a}_1^\dagger = \mathcal{R}\hat{a}_3^\dagger + \mathcal{T}\hat{a}_4^\dagger, \tag{1.7}$$

$$\hat{a}_2^\dagger = \mathcal{T}\hat{a}_3^\dagger - \mathcal{R}\hat{a}_4^\dagger. \tag{1.8}$$

The output state following the beamsplitter transformation is found by applying the creation operator of one of the input modes to the vacuum state $|0\rangle$,

$$|\psi\rangle = \hat{a}_1^\dagger|0\rangle = (\mathcal{R}\hat{a}_3^\dagger + \mathcal{T}\hat{a}_4^\dagger)|0\rangle, \tag{1.9}$$

$$|\psi\rangle = \hat{a}_2^\dagger|0\rangle = (\mathcal{T}\hat{a}_3^\dagger - \mathcal{R}\hat{a}_4^\dagger)|0\rangle. \tag{1.10}$$

If two single photons are input to the beamsplitter, one in each input mode, the output state is formed as:

$$|\psi\rangle = \hat{a}_1^\dagger \hat{a}_2^\dagger |0\rangle, \tag{1.11}$$

$$= (\mathcal{R}\hat{a}_3^\dagger + \mathcal{T}\hat{a}_4^\dagger)(\mathcal{T}\hat{a}_3^\dagger - \mathcal{R}\hat{a}_4^\dagger)|0\rangle. \tag{1.12}$$

$$= \mathcal{RT}\hat{a}_3^\dagger \hat{a}_3^\dagger + \mathcal{T}^2 \hat{a}_4^\dagger \hat{a}_3^\dagger - \mathcal{R}^2 \hat{a}_3^\dagger \hat{a}_4^\dagger - \mathcal{RT} \hat{a}_4^\dagger \hat{a}_4^\dagger |0\rangle. \tag{1.13}$$

For the balanced beam-splitter $\mathcal{R} = \mathcal{T} = 1/\sqrt{2}$, and as the operators $\hat{a}_3^\dagger$ and $\hat{a}_4^\dagger$ commute $\left(\left[\hat{a}_3^\dagger, \hat{a}_4^\dagger\right] = 0\right)$, the resulting output sate is,

$$|\psi\rangle = \mathcal{RT}\left(\hat{a}_3^\dagger \hat{a}_3^\dagger - \hat{a}_4^\dagger \hat{a}_4^\dagger\right)|0\rangle. \tag{1.14}$$

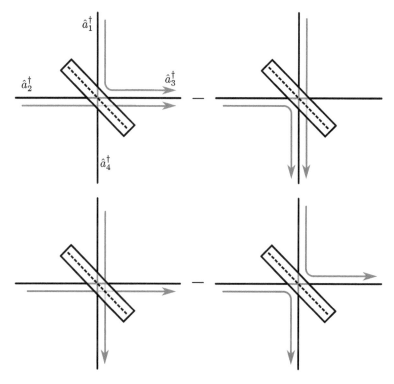

Fig. 1.2 Illustration of the four superposed terms of Eq. 1.13 corresponding to the interference of two single photons at a 50:50 beamsplitter. *Top panel* Both photons output the beam-splitter in the same mode, these two output states are distinguishable from one another. *Bottom panel* Either both photons are transmitted or reflected, if the photons themselves are indistinguishable then both of these two results are indistinguishable up to a global phase factor acquired by reflection. This results in destructive interference between these two states, leading to a superposition of only the two states in the *top panel*

1.1 What Is a Photon?

Using $\mathcal{R}^2 - \mathcal{T}^2 = 0$, $|\psi\rangle$ can be renormalised as,

$$|\psi\rangle = \frac{1}{\sqrt{2}} \left(\hat{a}_3^\dagger \hat{a}_3^\dagger - \hat{a}_4^\dagger \hat{a}_4^\dagger \right) |0\rangle. \tag{1.15}$$

The output state is still a linear superposition, but now the only terms remaining are those in which both photons exit in the same spatial mode. This is known as Hong-Ou-Mandel interference, and will only occur if the two input photons are in pure indistinguishable states [9, 26]. For indistinguishable inputs the terms corresponding to either both photons transmitted or reflected destructively interfere due to the π-phase difference acquired on reflection, as shown in Fig. 1.2.

HOM interference is a purely quantum mechanical effect and is key to the operation of quantum logic gates in the KLM implementation of LOQC as well as schemes for generating photonic cluster states [15, 30]. Additionally it is also an incredibly useful technique for experimentally determining the indistinguishability of two single photon states [31–33]. A controlled amount of distinguishing information can be introduced, for example in the temporal [9, 31, 34] or polarisation domain, which is then varied whilst monitoring the output modes of the beamsplitter. Only for identical initial states will photons always exit in the same mode. This technique provides a complete indication of the suitability of photons for quantum information tasks.

1.2 Methods of Single Photon Generation

There are two broad classes of single photon generation, single emitters consisting of one and only one source of emission such as an atom or ion, and multiple emitter sources which are most commonly based on non-linear optical processes. In the latter, emission occurs due to the coherent action of a number emitters which must be summed in phase to achieve a non-zero output. These sources based on non-linear optics produce photons in pairs, one of these photons can then detected to herald the presence of the remaining photon. Hence, these sources are termed heralded single photon sources.

1.2.1 Single Emitter Sources

The first of the two broad classes of single photon sources are those termed single emitter sources. In this class of source, single photons are emitted from a single discrete quantum system such as an atom, ion or quantum dot. As there is only a single emitter present, upon excitation by an external driving field, only a single photon will be emitted on de-excitation [35]. In order to ensure that the properties of photons emitted consecutively are identical (frequency, spatial distribution etc.), the environment surrounding the emitter must be carefully controlled. This is typically accomplished through the use of a cavity.

These systems can approach deterministic operation, provided that excitation mechanism has a high probability of success (optical pumping with a narrow linewidth laser), but are commonly limited by the collection efficiency as photons are emitted on a spherical surface centred on the emitter. This can be alleviated by using a cavity, in which case by coupling the emitter to a cavity mode, photons are then emitted into a well-defined spatial mode that is easier to collect. However, this is technically challenging. In addition to this, the vast majority of single emitter based sources require the use of vacuum systems, cryogenic equipment to reduce the coupling between the emitter and surrounding environment and prevent decoherence and laser cooling schemes requiring multiple carefully aligned laser systems. As a result of this, these systems are not user friendly or readily scalable and therefore limit their application for preparing a large number of photons simultaneously.

Trapped atoms are a natural choice for a single emitter source and there are a number of different atomic species with easily accessible electronic energy levels. When this atom is excited by the external driving field, one and only one photon will be emitted (provided that the energy level structure is chosen correctly) on de-excitation. The external driving field can then be pulsed to realise a steady stream of single photons from the system [36]. As there is little uncertainty in the energy of the electronic levels and the allowed cavity resonances, the linewidth of the emitted photons can be extremely narrow on the order of 10 MHz. This gives good coupling of the photons to other atomic transitions, such as in photonic storage schemes [37], but poor temporal purity.

The cavity trapped atom schemes can be realised in a number of different ways, such as using a magneto-optical trap to drop atoms one at a time through the cavity [38], trapping in an optical cavity mode [36] and optical lattice traps [39]. Alternatively, the atom can be replaced by an ion. The trapping is more stable and easier to achieve using radio-frequency dipole traps [40, 41].

A second approach is to engineer an artificial atomic system such as a quantum dot. Single photons are emitted from the recombination of electron-hole pairs that are produced when the dot is optically excited [42]. These sources suffer from some of the same problems as atomic base sources, in that photons are not emitted with a well-defined wavevector. Similarly, quantum dots can be constructed within microcavities formed using Bragg mirrors, to preferentially emit photons into a useful direction and constrain the optical frequency [43]. However, one of the largest problems associated with these sources is again scalability as these experiments must be carried out at cryogenic temperatures. It is difficult to achieve a high degree of spectral indistinguishability from two individual dots as the fabrication procedure often results in a series of dots with a distribution of transition energies. It has however been shown to be possible to tune the energy levels of the dots using an applied voltage to achieve sufficient spectral overlap to observe HOM interference [44].

The final implementation of single emitter sources consists of lattice defects in inorganic crystals, the most common being the nitrogen vacancy (NV) centre in diamond. The NV centre consists of a nitrogen atom substitution for one carbon atom and a neighbouring vacancy in the diamond lattice. The vacancy has an overall charge of -1 corresponding to the trapping of an electron within the defect. This electron can then be excited and de-excited to produce a stream of single photons [45, 46]. Unlike quantum dot sources, NV centres can be operated at room temperature, but the photons can be hard to extract due to the high refractive index of diamond, this can be improved by using diamond nanocrystals instead [47]. Whilst many of the single emitter sources can be operated at near deterministic performance, they are often limited by the inherent experimental complexity required to extract this performance. In contrast, heralded single photon sources are often easier to construct and maintain but at the disadvantage that they cannot naturally be operated deterministically.

1.2.2 Heralded Single Photon Sources

Amongst the most common single photon sources are those based on parametric photon pair generation, where one photon in the pair can be detected, and used to "herald" the presence of the remaining photon [8]. When an intense pump laser propagates through an optical medium, the nonlinear response ($\chi^{(2)}$ or $\chi^{(3)}$) leads to three-wave- and four-wave-mixing processes [48], through which pairs of photons are generated into two modes termed the signal and idler. For $\chi^{(2)}$ media this is called parametric downconversion (PDC) [49, 50], and for $\chi^{(3)}$ media four-wave mixing (FWM) [48, 51], the generated frequencies being determined by energy and momentum conservation.

The process is spontaneous and hence probabilistic, so there is no prior knowledge of when a generation event will occur [10]. But by using a pulsed pump, photon pairs can only be generated within certain time bins. However, in addition to producing a single pair, there is also the possibility of generating multiple pairs, which increases with the single pair probability. To limit the contribution of these multi-pair events, the source must be operated at a low probability of generating a single pair. As a result of this photon pairs can only be generated within specific time bins but the majority of these bins are empty [10]. Nevertheless, high brightness (photon counts per second) heralded single photon sources have been produced, most notably through PDC in $\chi^{(2)}$ crystals [52–54]. However, as a result of momentum conservation photons are emitted spatially around a ring centred on the pump wavevector. Pairs of photons must be spatially filtered to ensure collection in a single spatial mode. As a result of this, these photons can be hard to collect as the spatial emission pattern is not well matched to optical fibres [55]. These sources, whilst composed of relatively simple components, require careful alignment to maintain high performance operation.

An alternative approach is to make use of the FWM interaction in $\chi^{(3)}$ optical fibres [48]. The generation, collection, delivery and detection (fibre coupled detectors) can all be achieved in optical fibre, allowing for a truly integrated device in which the photons are emitted into a well-defined spatial mode and which is significantly easier to maintain for the end user [32, 56]. A number of similar approaches have been made using waveguide sources, for both PDC [57, 58] and FWM in optical fibres [51, 59–61], and in recent years FWM in chip-based waveguides such as silicon photonic crystal waveguides [62, 63], chalcogenide waveguide [64], and UV-written fused silica waveguides [65]. Although only the fibre-based source of McMillan et al. [56] could be described as a true integrated device.

An added complication of parametric photon pair generation is that due to energy and momentum conservation photon pairs are typically generated into a number of correlated spectral modes, rather than the ideal single spectral mode [66, 67]. This leads to a mixed state on heralding, which subsequently must be filtered to approach a single spectral mode, resulting in a reduced brightness from the source. However, generation into a single spectral mode has been demonstrated, for both PDC [31, 34] and FWM [68, 69] by exploiting the dispersion of the medium to match the group velocities of the pump, signal and idler fields, this is discussed in more detail in Sect. 2.3.

Typically in both PDC and FWM, the dispersion of the medium is dominated by the material. This places limits on the frequencies of the generated photon pairs. FWM in photonic crystal fibres (PCF) [56, 59, 60], where the waveguiding structure is formed from an array of low-index inclusions in the cladding, allows unprecedented control over the waveguide contribution of the dispersion. By tailoring the parameters of the cladding, the dispersion and propagation constants can be controlled, producing photon pairs at desired wavelengths, such as 800/1550 nm, in a single spectral mode [56, 70, 71].

FWM in optical fibres does have some limitations itself. The $\chi^{(3)}$ nonlinearity is several orders of magnitude smaller than $\chi^{(2)}$, resulting in lower overall count rates. However, the guided mode nature of the FWM process means a long interaction length can be used to compensate. The largest deficiency of FWM based sources is the often high levels of noise due to other nonlinear processes generating uncorrelated photons into similar spectral modes as the heralded idler photon. Again, thanks to the dispersion of the medium, these effects can be reduced by generating photon pairs far detuned from the pump wavelength where the majority of the noise is located. Alternatively, in birefringent optical fibres, cross-polarised FWM can be used to generate photons with an orthogonal polarisation state to the pump. Much of the noise will be generated co-polarised with the pump field and so polarisation filtering can be used to recover the photon pairs [72]. To date there have been a large number of sources utilising both PDC and FWM, where the heralded single photons display high levels of purity and indistinguishability, they are however all still probabilistic in nature.

1.3 Multiplexing: A Route to Deterministic Operation

As discussed, sources of heralded single photons based on PDC and FWM suffer from a fundamental limitation: The mechanism is spontaneous. One can only say that a single photon pair will be generated with some probability per pulse of the

1.3 Multiplexing: A Route to Deterministic Operation

laser, $P(1)$. In order to limit detrimental multi-pair emission from the source, $P(1)$ is typically constrained to around $P(1) = 0.01$. Hence the performance of the source is far from ideal deterministic operation. As a result of this, in order to deliver N_p heralded independent photons from N_p independent sources, requires waiting for all sources to fire simultaneously, the probability of which scales exponentially with the number of photons requested,

$$p(N_p) = P(1)^{N_p}, \qquad (1.16)$$

where $p(N_p)$ is the probability of delivering N_p photons simultaneously. This translates into a long waiting time for the delivery of N_p photons simultaneously. Consider a typical photon pair source pumped by a 76 MHz Titanium:Sapphire laser. The approximate waiting time t_{wait} for all N_p sources to fire simultaneously is,

$$t_{\text{wait}} = \left(\frac{1}{p(N_p) \cdot R_p} \right), \qquad (1.17)$$

where R_p is the repetition rate of the laser train, this is illustrated in Fig. 1.3.

Even for a modest number of heralded photons ($N_p = 8$), the waiting time is on the order of 400 years. Ultimately, many more than 8 single photon qubits may be required by a future device. Developing a source that is capable of delivering single photons in a deterministic manner is therefore a key research task.

To overcome the fundamental limit of probabilistic photon pair sources, Migdall et al. [10] proposed that an array of PDC sources could be combined within an active optical switching network. In doing so, a photon generated by any one of the sources

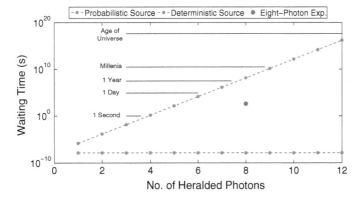

Fig. 1.3 Approximate waiting times to deliver N_p independent photons from N_p independent probabilistic sources. A single source pumped with $P(1) = 0.01$ at 76 MHz (*blue*). For even modest numbers of photons the waiting time extends towards the age of universe. Where as for a deterministic source (*green*), every one fires on every laser pulse and the waiting time does not scale with N_p but remains fixed at the time period of the laser pulse train. Example of eight-photon state generation using 4 PDC sources (*red*) with $P(1) = 0.058$ per source at 76 MHz [11]

may be routed to a common output port. Heralded photon pair sources are particularly suited to this scheme as the results of any heralding detection event can be used to feed forward to set the state of the switches to route the corresponding photon to the output. With a sufficient number of individual sources, a deterministic multiplexed system can be constructed, where each source can be operated at a level where the multi-pair emission is low.

A number of different multiplexing schemes have been proposed in the spatial [10, 73], temporal [74–79] and spectral domains [80]. In the spatial domain, as considered by Migdall et al. a number of individual PDC sources can be used or multiple passes through the same pair-generation medium, in an effort to reduce the amount of components required. Alternatively, in bulk crystal PDC pairs of photons can be collected from different spatial modes allowing a single crystal to be used instead of a number of distinct sources [10]. Collection is however difficult and potentially lossy.

In the temporal domain, several different procedures have been explored including the use of storage loops or optical cavities and delay lines. In the foremost case using storage loops or optical cavities, when a photon pair is heralded, the heralding signal is used to activate a switch which routes the heralded photon into an optical storage loop. The photon can be held in the loop until it is required at a later time. This scheme was extended by Jeffery et al. [75], by dividing the pump pulse into a number of time bins, such that a short burst of pump pulses, with each burst separated by the repetition rate of the laser, are used to pump the medium. The time stamping of the heralding signal allows photons from specific sub-pulses within the burst to be routed into the storage loop [75].

Other schemes in the temporal domain include generating photon pairs and then routing the heralded photon through a number of delay lines, whose lengths are binary multiples of each other [78]. By comparing the heralding time-stamp to an external oscillator, a binary combination of delay can be inserted in the heralded arm with fast optical switches, such that the heralded photon is delivered at a specific point in time relative to the external clock forming a pseudo-on-demand single photon source.

The initial storage loop studies described above only ever considered switching in one event per series of pulses. However due to loss in the components, the stored photon may be lost. Jeffery et al. [75] demonstrated that allowing the cavity to be replenished up to two times within a burst of pump pulses produce a significant improvement. The number of times the cavity can be refilled is determined by the switching speed of the optical switches. Rohde et al. [79] and work undertaken by myself in the course of this thesis (see Sect. 3.4), have numerically demonstrated that if the storage loop is continually replenished whenever a new photon is heralded, the probability of success is greatly improved. This can be achieved by not using a burst of pulses as in the work of Jeffery et al. [75], but rather just using a series of pulses from the laser train. Provided that both the heralding detector dead time is less than the laser repetition rate, and also that the repetition rate of the laser is less than the switching time then the most recent photon can be captured. After every m pulses, the loop is either emptied extracting the stored photon if there was no pair generation of the mth pulse or the heralded photon from the mth pulse is used.

1.3 Multiplexing: A Route to Deterministic Operation

In all of the temporal schemes discussed above, an increase in the probability of delivering a heralded single photon can be achieved, but at the expense of the rate at which photons can be delivered. An alternative avenue, albeit similar in principle to some of the storage loop procedures, is to store the heralded single photon in a quantum memory with controllable read-out and delay [37]. This would allow many individual sources to be synchronised, so that when several photons are requested the memories can be read out delivering the required single photons to the experiment or device.

A passive temporal multiplexing scheme has been discussed by Broome et al. [81] whereby working at a high repetition rate, the effect of multi-photon emission from the source can be reduced. This allows the signal-to-noise of heralded single photons to be increased without affecting the brightness of the source. A number of groups have recently demonstrated high repetition rate photon pair sources [82–84] where this effect has been successfully measured through the degree of second order coherence [84].

Previous theoretical considerations of spatial multiplexing schemes utilising PDC have shown that by multiplexing several non-deterministic sources together, deterministic operation can be approached [85]. Christ and Silberhorn [85] examined the impact of multi-photon emission and multiple spectral modes on the performance of spatially multiplexed systems. They found that for a system composed of 17 identical PDC sources, utilising perfect photon-number resolving detectors (PNR), a single photon per pulse delivery probability of >99.9% could be achieved. In this ideal case of lossless components, the overall probability of successfully delivering a single photon from the multiplexed system, $p(\text{success})$, is just the probability that all the sources do not fail to generate a photon pair, i.e.,

$$p(\text{success}) = 1 - (1 - P_1)^{N_s}, \quad (1.18)$$

where N_s is the number of individual sources multiplexed together. For a photon pair state described by Bose-Einstein (thermal) statistics the maximum value of $P_1 = 0.25$, at a mean photon number per pulse $\bar{n} = 1$. Therefore to achieve $p(\text{success}) > 99.9\%$, requires $N_s \geq 17$. thermal distribution of photon number, pumping at the peak value of $p_{(1)} = 0.25$ corresponding to a mean photon number $\bar{n} = 1$, a value of $p(\text{success}) > 99.9\%$ is achievable.

However this is for an idealised set of circumstances. Commercially available single photon detectors and optical switches will have some inherent inefficiency. Current silicon single-photon avalanche photo-diodes (APD) have peak efficiencies in the region of 70% and are binary in operation, providing no information about the number of photons that strike the detector element. Photon-number resolving detectors with high detection efficiency around 800 nm are available, but the timing jitter is often large and the speed slow [86]. Commercial fibre coupled optical switches from the telecommunications industry operating at 1550 nm have efficiencies in the range of 70 − 80% with repetition rates around 1 MHz. The switch bandwidth and detector response time limit the laser repetition rate that can be used and therefore the maximum rate at which photons can be heralded and routed through the switching

Table 1.1 Multiplexing scheme performance comparison. Medium = pair generation process and nonlinear material. Multiplexed = implemented multiplexing scheme (S = spatial, T = temporal). Depth = number of multiplexed bins, or stages used. Pure State = spectral engineering carried out. $g^{(2)}(0)$ = second order coherence measurement. C.R. = count rate or brightness. Rep. rate = maximum switching rate

Year	Paper	Medium	Multiplexed	Depth	$g^{(2)}(0)$	C.R.	Rep. Rate
2002	Pittman [74]	PDC in BBO	Storage loop (T)	5	–	2 C/s	75 MHz
2004	Jeffery [75]	PDC in BBO	Storage loop (T)	–	–	–	10–50 kHz
2011	Ma [87]	PDC in BBO	Log-Tree (S)	4	0.08	714 C/s	~15 MHz
2011	Broome [81]	PDC in BBO	Pulse doubling (T)	2	–	40 C/s/mW	152 MHz
2013	Collins [88]	FWM in PhCW	Log-Tree (S)	2	~0.19	1–5 C/s	1 MHz
2014	Meany [89]	PDC in PPLNW	Log-Tree (S)	4	–	60–70 C/s	1 MHz
2015	Kaneda [90]	PDC in BiBO	Storage loop (T)	35	0.479	19.3 kC/s	50 kHz
2016	Mendoza [91]	PDC in PPLN	Tree + Delay (S and T)	2 and 4	–	140 C/s	500 KHz
2016	Xiong [92]	FWM in silicon nanowire	Delay lines (T)	4	–	600 C/s	10 MHz

network. Ultimately, multiplexed sources will be limited by the speed and efficiency of both the heralding detectors and switches.

To date there have been relatively few implementations of multiplexed systems, the performance of these systems is compared in Table 1.1, although the field is now growing rapidly. The first implementations of these multiplexing schemes were based on PDC in a free space set-up, using electro-optic modulators (EOM) as switches. The cost and experimental complexity of these schemes limit their scalability, although high repetition rates can be achieved due to the fast switching speeds of the EOMs. The more recent developments of multiplexed systems (2013–2015) have moved towards a fibre integrated switching network. These switches are significantly cheaper, easier to operate, require no alignment with a small footprint, and are therefore more suitable to being scaled up. A fibre based scheme also benefits from the ability to easily incorporate low loss optical delays of a chosen length using inexpensive commercial optical fibres. The cost of this however is slower switching times and hence lower repetition rates and also loss in the switch.

All of the multiplexed systems outline in Table 1.1 are not truly integrated devices. Whilst the pair generation is generally accomplished in one architecture, the photons are then extracted into the fibre switch network, with no permanent bridge between the two. Additionally no system has been demonstrated that is capable of generating and then multiplexing heralded single photons in spectrally engineered pure states.

An all-optical fibre based multiplexed system is an attractive prospect, as it should be easier to maintain and could be packaged and made portable. The use of PCF also allows the possibility of spectrally engineering the heralded single photons such that they are in a pure state, whilst also maintaining the benefits of a fibre based system. These fibre based sources are well suited for multiplexing schemes as photons can be extracted from the source into the switching network with lower losses, with a permanent mechanical link between the two. Additionally, the wealth of technology from the telecommunications industry can be translated across into these systems. This thesis presents the design, fabrication and characterisation of a multiplexed source of spectrally engineered heralded single photons in a single integrated fibre architecture.

1.4 Thesis Outline

This thesis is presented in seven chapters. Following this introduction, in Chap. 2, the quantum theory of FWM in PCF is laid out, and the ability to herald single photons in pure states is discussed. Chapter 3 develops a numerical modelling technique by which these multiplexed systems can be optimised given the inevitable loss in the constituent components. Chapter 4 provides an overview of the design, fabrication and classical characterisation of the PCF used for pair generation. The fifth chapter extends the characterisation by measuring the joint spectral probability distribution of the two-photon state via the technique of stimulated emission tomography. Also in Chap. 5, variations in the PCF due to fabrication tolerances and their effect on the purity of the heralded state are investigated. Chapter 6 provides an in-depth discussion of the characterisation of the generated photons, including measurements of the degree of second order coherence to determine the purity of the heralded state. Finally in Chap. 7, the conclusions of this passage of work are laid out and an outlook on future work presented.

References

1. J.F. Clauser, Experimental distinction between the quantum and classical field-theoretic predictions for the photoelectric effect. Phys. Rev. D **9**, 853–860 (1974)
2. H.J. Kimble, M. Dagenais, L. Mandel, Photon antibunching in resonance Fluorescence. Phys. Rev. Lett. **39**, 691–695 (1977)
3. D.F. Walls, Evidence for the quantum nature of light. Nature **280**, 451–454 (1979)
4. J.F. Clauser, A. Shimony, Bell's theorem. Experimental tests and implications. Rep. Prog. Phys. **41**, 1881 (1978)
5. A. Aspect, P. Grangier, G. Roger, Experimental tests of realistic local theories via Bell's theorem. Phys. Rev. Lett. **47**, 460–463 (1981)
6. T.H. Maiman, Stimulated Optical Radiation in Ruby. Nature **187**, 493–494 (1960)
7. D.C. Burnham, D.L. Weinberg, Observation of simultaneity in parametric production of optical photon pairs. Phys. Rev. Lett. **25**, 84–87 (1970)
8. C.K. Hong, L. Mandel, Experimental realization of a localized one-photon state. Phys. Rev. Lett. **56**, 58–60 (1986)
9. C.K. Hong, Z.Y. Ou, L. Mandel, Measurement of subpicosecond time intervals between two photons by interference. Phys. Rev. Lett. **59**, 2044–2046 (1987)
10. A.L. Migdall, D. Branning, S. Castelletto, Tailoring single-photon and multiphoton probabilities of a single-photon on-demand source. Phys. Rev. A **66**, 053805 (2002)
11. X.-C. Yao, T.-X. Wang, P. Xu, H. Lu, G.-S. Pan, X.-H. Bao, C.-Z. Peng, C.-Y. Lu, Y.-A. Chen, J.-W. Pan, Observation of eight-photon entanglement. Nat. Photon **6**, 225–228 (2012)
12. C.H. Bennett, D.P. DiVincenzo, Quantum information and computation. Nature **404**, 247–255 (2000)
13. J.L. O'Brien, Optical quantum computing. Science **318**, 1567–1570 (2007)
14. E. Knill, R. Laflamme, G.J. Milburn, A scheme for efficient quantum computation with linear optics. Nature **409**, 46–52 (2001)
15. D.E. Browne, T. Rudolph, Resource-efficient linear optical quantum computation. Phys. Rev. Lett. **95**, 010501 (2005)
16. P. Kok, W.J. Munro, K. Nemoto, T.C. Ralph, J.P. Dowling, G.J. Milburn, Linear optical quantum computing with photonic qubits. Rev. Mod. Phys. **79**, 135–174 (2007)
17. T. Schmitt-Manderbach, H. Weier, M. Fürst, R. Ursin, F. Tiefenbacher, T. Scheidl, J. Perdigues, Z. Sodnik, C. Kurtsiefer, J.G. Rarity, A. Zeilinger, H. Weinfurter, Experimental demonstration of free-space decoy-state quantum key distribution over 144 km. Phys. Rev. Lett. **98**, 010504 (2007)
18. H.-K. Lo, M. Curty, K. Tamaki, Secure quantum key distribution. Nat. Photon **8**, 595–604 (2014)
19. Networked quantum information technologies, http://nqit.ox.ac.uk/. Accessed 20 Dec 2015
20. A. Aspect, P. Grangier, G. Roger, Experimental realization of Einstein-Podolsky-Rosen-Bohm Gedankenexperiment: a new violation of Bell's inequalities. Phys. Rev. Lett. **49**, 91–94 (1982)
21. Z.Y. Ou, L. Mandel, Violation of Bell's inequality and classical probability in a two-photon correlation experiment. Phys. Rev. Lett. **61**, 50–53 (1988)
22. P.R. Tapster, J.G. Rarity, P.C.M. Owens, Violation of Bell's inequality over 4 km of optical fiber. Phys. Rev. Lett. **73**, 1923–1926 (1994)
23. A. Aspect, Bell's inequality test: more ideal than ever. Nature **398**, 189–190 (1999)

References

24. T.B. Pittman, J.D. Franson, Violation of Bell's inequality with photons from independent sources. Phys. Rev. Lett. **90**, 240401 (2003)
25. A. Migdall, S.G. Polyakov, J. Fan, J.C. Beinfang (eds.), Single-photon generation and detection, in *Experimental Methods in the Physical Sciences*, vol. 45 (Elsevier, 2013)
26. R. Loudon, *The Quantum Theory of Light*, vol. 1, 2nd edn. (Oxford University Press, Oxford, 1973)
27. L. Mandel, E. Wolf, *Optical Coherence and Quantum Optics*, 2nd edn. (Cambridge Unviersity Press, Cambridge, 1995)
28. A.B. U'Ren, Multi-photon state engineering for quantum information processing applications. Ph.D. thesis, University of Rochester, 2004
29. P.J. Mosley, Generation of heralded single photons in pure quantum states. Ph.D. thesis, University of Oxford, 2007
30. B.A. Bell, D.A. Herrera-Martí, M.S. Tame, D. Markham, W.J. Wadsworth, J.G. Rarity, Experimental demonstration of a graph state quantum error-correction code. Nat. Commun. **5** (2014)
31. P.J. Mosley, J.S. Lundeen, B.J. Smith, I.A. Walmsley, Conditional preparation of single photons using parametric downconversion: a recipe for purity. New J. Phys. **10**, 093011 (2008)
32. A. McMillan, Development of an all-fibre source of heralded single photons. Ph.D. thesis, University of Bath, 2011
33. G. Harder, V. Ansari, B. Brecht, T. Dirmeier, C. Marquardt, C. Silberhorn, An optimized photon pair source for quantum circuits. Opt. Express **21**, 13975–13985 (2013)
34. P.J. Mosley, J.S. Lundeen, B.J. Smith, P. Wasylczyk, A.B. U'Ren, C. Silberhorn, I.A. Walmsley, Heralded generation of ultrafast single photons in pure quantum states. Phys. Rev. Lett. **100**, 133601 (2008)
35. B. Lounis, M. Orrit, Single-photon sources. Rep. Prog. Phys. **68**, 1129 (2005)
36. J. McKeever, A. Boca, A.D. Boozer, R. Miller, J.R. Buck, A. Kuzmich, H.J. Kimble, Deterministic generation of single photons from one atom trapped in a cavity. Science **303**, 1992–1994 (2004)
37. J. Nunn, N.K. Langford, W.S. Kolthammer, T.F.M. Champion, M.R. Sprague, P.S. Michelberger, X.-M. Jin, D.G. England, I.A. Walmsley, Enhancing multiphoton rates with quantum memories. Phys. Rev. Lett. **110**, 133601 (2013)
38. A. Kuhn, M. Hennrich, G. Rempe, Deterministic single-photon source for distributed quantum networking. Phys. Rev. Lett. **89**, 067901 (2002)
39. M. Hijlkema, B. Weber, H.P. Specht, S.C. Webster, A. Kuhn, G. Rempe, A single-photon server with just one atom. Nat. Phys. **3**, 253–255 (2007)
40. M. Keller, B. Lange, K. Hayasaka, W. Lange, H. Walther, Continuous generation of single photons with controlled waveform in an ion-trap cavity system. Nature **431**, 1075–1078 (2004)
41. H.G. Barros, A. Stute, T.E. Northup, C. Russo, P.O. Schmidt, R. Blatt, Deterministic single-photon source from a single ion. New J. Phys. **11**, 103004 (2009)
42. A.J. Shields, Semiconductor quantum light sources. Nat. Photon **1**, 215–223 (2007)
43. J.C. Loredo, N.A. Zakaria, N. Somaschi, C. Anton, L. de Santis, V. Giesz, T. Grange, M.A. Broome, O. Gazzano, G. Coppola, I. Sagnes, A. Lemaitre, A. Auffeves, P. Senellart, M.P. Almeida, A.G. White, Scalable performance in solid-state single-photon sources. Optica **3**, 433–440 (2016)
44. R.B. Patel, A.J. Bennett, I. Farrer, C.A. Nicoll, D.A. Ritchie, A.J. Shields, Two-photon interference of the emission from electrically tunable remote quantum dots. Nat. Photon **4**, 632–635 (2010)
45. A. Beveratos, S. Kühn, R. Brouri, T. Gacoin, J.-P. Poizat, P. Grangier, Room temperature stable single-photon source. Eur. Phys. J. D **18**, 191–196 (2002)
46. D. Englund, B. Shields, K. Rivoire, F. Hatami, J. Vučković, H. Park, M.D. Lukin, Deterministic coupling of a single nitrogen vacancy center to a photonic crystal cavity. Nano Lett. **10**, 3922–3926 (2010). pMID: 20825160
47. A. Beveratos, R. Brouri, T. Gacoin, A. Villing, J.-P. Poizat, P. Grangier, Single photon quantum cryptography. Phys. Rev. Lett. **89**, 187901 (2002)
48. R. Boyd, *Non-Linear Optics*, 3rd edn. (Academic Press, Burlington, 2008)

49. S. Friberg, C.K. Hong, L. Mandel, Measurement of time delays in the parametric production of photon pairs. Phys. Rev. Lett. **54**, 2011–2013 (1985)
50. C.K. Hong, L. Mandel, Theory of parametric frequency down conversion of light. Phys. Rev. A **31**, 2409–2418 (1985)
51. J.E. Sharping, M. Fiorentino, P. Kumar, Observation of twin-beam-type quantum correlation in optical fiber. Opt. Lett. **26**, 367–369 (2001)
52. C. Kurtsiefer, M. Oberparleiter, H. Weinfurter, High-efficiency entangled photon pair collection in type-II parametric fluorescence. Phys. Rev. A **64**, 023802 (2001)
53. A.B. U'Ren, C. Silberhorn, K. Banaszek, I.A. Walmsley, Efficient conditional preparation of high-fidelity single photon states for fiber-optic quantum networks. Phys. Rev. Lett. **93**, 093601 (2004)
54. T. Pittman, B. Jacobs, J. Franson, Heralding single photons from pulsed parametric down-conversion. Opt. Commun. **246**, 545–550 (2005)
55. F.A. Bovino, P. Varisco, A.M. Colla, G. Castagnoli, G.D. Giuseppe, A.V. Sergienko, Effective fiber-coupling of entangled photons for quantum communication. Opt. Commun. **227**, 343–348 (2003)
56. A.R. McMillan, J. Fulconis, M. Halder, C. Xiong, J.G. Rarity, W.J. Wadsworth, Narrowband high-fidelity all-fibre source of heralded single photons at 1570 nm. Opt. Express **17**, 6156–6165 (2009)
57. M.G. Raymer, J. Noh, K. Banaszek, I.A. Walmsley, Pure-state single-photon wave-packet generation by parametric down-conversion in a distributed microcavity. Phys. Rev. A **72**, 023825 (2005)
58. J. Chen, A.J. Pearlman, A. Ling, J. Fan, A. Migdall, A versatile waveguide source of photon pairs for chip-scale quantum information processing. Opt. Express **17**, 6727–6740 (2009)
59. J. Fan, A. Migdall, L.J. Wang, Efficient generation of correlated photon pairs in a microstructure fiber. Opt. Lett. **30**, 3368–3370 (2005)
60. J. Rarity, J. Fulconis, J. Duligall, W. Wadsworth, P. Russell, Photonic crystal fiber source of correlated photon pairs. Opt. Express **13**, 534–544 (2005)
61. J. Fulconis, O. Alibart, W. Wadsworth, P. Russell, J. Rarity, High brightness single mode source of correlated photon pairs using a photonic crystal fiber. Opt. Express **13**, 7572–7582 (2005)
62. C. Xiong, C. Monat, M. Collins, L. Tranchant, D. Petiteau, A. Clark, C. Grillet, G. Marshall, M. Steel, J. Li, L. O'Faolain, T. Krauss, B. Eggleton, Characteristics of correlated photon pairs generated in ultracompact silicon slow-light photonic crystal waveguides. IEEE J. Sel. Top. Quantum Electron. **18**, 1676–1683 (2012)
63. C. Xiong, M. Collins, M. Steel, T. Krauss, B. Eggleton, A. Clark, Photonic crystal waveguide sources of photons for quantum communication applications. IEEE J. Sel. Top. Quantum Electron. **21**, 1–10 (2015)
64. M.J. Collins, A.S. Clark, J. He, D.-Y. Choi, R.J. Williams, A.C. Judge, S.J. Madden, M.J. Withford, M.J. Steel, B. Luther-Davies, C. Xiong, B.J. Eggleton, Low Raman-noise correlated photon-pair generation in a dispersion-engineered chalcogenide As2S3 planar waveguide. Opt. Lett. **37**, 3393–3395 (2012)
65. J.B. Spring, P.S. Salter, B.J. Metcalf, P.C. Humphreys, M. Moore, N. Thomas-Peter, M. Barbieri, X.-M. Jin, N.K. Langford, W.S. Kolthammer, M.J. Booth, I.A. Walmsley, On-chip low loss heralded source of pure single photons. Opt. Express **21**, 13522–13532 (2013)
66. W.P. Grice, I.A. Walmsley, Spectral information and distinguishability in type-II down-conversion with a broadband pump. Phys. Rev. A **56**, 1627–1634 (1997)
67. W.P. Grice, A.B. U'Ren, I.A. Walmsley, Eliminating frequency and space-time correlations in multiphoton states. Phys. Rev. A **64**, 063815 (2001)
68. O. Cohen, J.S. Lundeen, B.J. Smith, G. Puentes, P.J. Mosley, I.A. Walmsley, Tailored Photon-Pair Generation in Optical Fibers. Phys. Rev. Lett. **102**, 123603 (2009)
69. C. Söller, O. Cohen, B.J. Smith, I.A. Walmsley, C. Silberhorn, High-performance single-photon generation with commercial-grade optical fiber. Phys. Rev. A **83**, 031806 (2011)
70. X. Li, P.L. Voss, J.E. Sharping, P. Kumar, Optical-fiber source of polarization-entangled photons in the 1550 nm telecom band. Phys. Rev. Lett. **94**, 053601 (2005)

References

71. C. Söller, B. Brecht, P.J. Mosley, L.Y. Zang, A. Podlipensky, N.Y. Joly, P.S.J. Russell, C. Silberhorn, Bridging visible and telecom wavelengths with a single-mode broadband photon pair source. Phys. Rev. A **81**, 031801 (2010)
72. B.J. Smith, P. Mahou, O. Cohen, J.S. Lundeen, I.A. Walmsley, Photon pair generation in birefringent optical fibers. Opt. Express **17**, 23589–23602 (2009)
73. J.H. Shapiro, F.N. Wong, On-demand single-photon generation using a modular array of parametric downconverters with electro-optic polarization controls. Opt. Lett. **32**, 2698–2700 (2007)
74. T.B. Pittman, B.C. Jacobs, J.D. Franson, Single photons on pseudodemand from stored parametric down-conversion. Phys. Rev. A **66**, 042303 (2002)
75. E. Jeffrey, N.A. Peters, P.G. Kwiat, Towards a periodic deterministic source of arbitrary single-photon states. New J. Phys. **6**, 100 (2004)
76. K.T. McCusker, P.G. Kwiat, Efficient optical quantum state engineering. Phys. Rev. Lett. **103**, 163602 (2009)
77. J. Mower, D. Englund, Efficient generation of single and entangled photons on a silicon photonic integrated chip. Phys. Rev. A **84**, 052326 (2011)
78. C.T. Schmiegelow, M.A. Larotonda, Multiplexing photons with a binary division strategy. Appl. Phys. B **116**, 447–454 (2013)
79. P.P. Rohde, L.G. Helt, M.J. Steel, A. Gilchrist, Multiplexed single-photon state preparation using a fibre-loop architecture, ArXiv e-prints (2015)
80. R. Kumar, J.R. Ong, J. Recchio, K. Srinivasan, S. Mookherjea, Spectrally multiplexed and tunable-wavelength photon pairs at 1.55 μm from a silicon coupled-resonator optical waveguide. Opt. Lett. **38**, 2969–2971 (2013)
81. M.A. Broome, M.P. Almeida, A. Fedrizzi, A.G. White, Reducing multi-photon rates in pulsed down-conversion by temporal multiplexing. Opt. Express **19**, 22698–22708 (2011)
82. O.J. Morris, R.J. Francis-Jones, K.G. Wilcox, A.C. Tropper, P.J. Mosley, Photon-pair generation in photonic crystal fibre with a 1.5GHz modelocked VECSEL. Opt. Commun. **327**, 39–44 (2014). Special Issue on Nonlinear Quantum Photonics
83. R.-B. Jin, R. Shimizu, I. Morohashi, K. Wakui, M. Takeoka, S. Izumi, T. Sakamoto, M. Fujiwara, T. Yamashita, S. Miki, H. Terai, Z. Wang, M. Sasaki, Efficient generation of twin photons at telecom wavelengths with 2.5 GHz repetition-rate-tunable comb laser. Sci. Rep. **4**, 7468 EP (2014)
84. L.A. Ngah, O. Alibart, L. Labonté, V. D'Auria, S. Tanzilli, Ultra-fast heralded single photon source based on telecom technology. Laser Photonics Rev. **9**, L1–L5 (2015)
85. A. Christ, C. Silberhorn, Limits on the deterministic creation of pure single-photon states using parametric down-conversion. Phys. Rev. A **85**, 023829 (2012)
86. T.D. Ladd, F. Jelezko, R. Laflamme, Y. Nakamura, C. Monroe, J.L. O'Brien, Quantum computers. Nature **464**, 45–53 (2010)
87. X.-S. Ma, S. Zotter, J. Kofler, T. Jennewein, A. Zeilinger, Experimental generation of single photons via active multiplexing. Phys. Rev. A **83**, 043814 (2011)
88. M.J. Collins, C. Xiong, I.H. Rey, T.D. Vo, J. He, S. Shahnia, C. Reardon, T.F. Krauss, M.J. Steel, A.S. Clark, B.J. Eggleton, Integrated spatial multiplexing of heralded single-photon sources. Nat. Commun. **4** (2013)
89. T. Meany, L.A. Ngah, M.J. Collins, A.S. Clark, R.J. Williams, B.J. Eggleton, M.J. Steel, M.J. Withford, O. Alibart, S. Tanzilli, Hybrid photonic circuit for multiplexed heralded single photons. Laser Photonics Rev. **8**, L42–L46 (2014)
90. F. Kaneda, B.G. Christensen, J.J. Wong, H.S. Park, K.T. McCusker, P.G. Kwiat, Time-multiplexed heralded single-photon source. Optica **2**, 1010–1013 (2015)
91. G.J. Mendoza, R. Santagati, J. Munns, E. Hemsley, M. Piekarek, E. Martín-López, G.D. Marshall, D. Bonneau, M.G. Thompson, J.L. O'Brien, Active temporal and spatial multiplexing of photons. Optica **3**, 127–132 (2016)
92. C. Xiong, X. Zhang, Z. Liu, M. J. Collins, A. Mahendra, L.G. Helt, M.J. Steel, D.Y. Choi, C.J. Chae, P.H.W. Leong, B.J. Eggleton, Active temporal multiplexing of indistinguishable heralded single photons. Nat. Commun. **7** (2016)

Chapter 2
Photon Pair Generation via Four-Wave Mixing in Photonic Crystal Fibres

2.1 Overview

The four-wave mixing process is well known in classical non-linear optics but has also been studied extensively within the framework of quantum theory, due to its usefulness as a source of heralded single photons. A short consideration of the theory is made here, as knowledge of the form of the two-photon state is required to allow us to engineer particularly useful quantum states from the interaction. In this thesis we are primarily concerned with the generation of single, indistinguishable photons in pure quantum states.

FWM is a third-order nonlinear optical process that occurs in all optical materials in which an intense pump field propagates, generating new frequencies of light in the process. Due to the electric field component, $\mathbf{E}(\vec{r}, t)$ of the pump, driving the bound electrons, the material acquires a polarisation $\mathbf{P}(\vec{r}, t)$ given by:

$$\mathbf{P} = \varepsilon_0 (\chi^{(1)} \mathbf{E} + \chi^{(2)} \mathbf{EE} + \chi^{(3)} \mathbf{EEE} + \cdots), \tag{2.1}$$

where $\chi^{(i)}$ is the i'th order dielectric susceptibility coefficient. In an optical fibre, due to the isotropic nature of the silica glass, there is no $\chi^{(2)}$ non-linearity present, and so the dominant non-linear processes arise through the $\chi^{(3)}$ non-linearity.

FWM is one of the resulting processes due to the 3rd order, $\chi^{(3)}$, dielectric susceptibility, which in general allows the coupling of four fields via the polarisation of the medium. This allows two pump fields at frequency ω_p to lose energy, which is re-distributed into two fields termed the signal at a higher frequency, ω_s, and the idler at lower frequency ω_i. In general the pump fields may be non-degenerate in frequency, however within the context of this thesis only degenerate pump fields produced by the same intense pump laser are considered.

Two different forms of the FWM interaction may occur, either stimulated or spontaneous FWM, both of which are used within this thesis. In the stimulated form of FWM, a seed beam is also injected into the material at either of the signal or idler frequencies. This seed field is amplified, along with the generation of the light in the other mode. If there is no seed present, then spontaneous FWM may occur, in which both the signal and idler modes are initially in the vacuum state, in this case the process is seeded by quantum vacuum fluctuations. In Chap. 4 stimulated FWM is used a diagnostic tool to determine the spectral distribution of the light generated in the signal mode. Whereas in Chap. 6, the spontaneous form is used for photon pair generation from which heralded single photons are produced.

For either stimulated or spontaneous FWM, the frequencies of the generated light are constrained by energy conservation,

$$2\omega_p = \omega_s + \omega_i, \tag{2.2}$$

and momentum conservation also known as phasematching,

$$\Delta k = 2k_p(\omega_p) - k_s(\omega_s) - k_i(\omega_i) - 2\gamma P, \tag{2.3}$$

where $k_j(\omega_j)$ are the propagation constants of the pump, signal and idler fields, and $2\gamma P$ is the phase shift of the pump due to self-phase modulation (SPM). Self-phase modulation is another $\chi^{(3)}$ nonlinear process in which due to the intense pump envelope, the pulse experiences an intensity-dependent phase shift. This results in the generation of new spectral components at the leading and trailing edges of the pulse envelope producing a chirped pulse. It can be shown that only if $\Delta k \approx 0$ will there be an appreciable probability of generating a photon pair. However due to the dispersion of the medium this condition can only be satisfied for certain frequencies.

This phasematching condition arises as a result of considering the propagation of the three fields at their respective phase velocities through the non-linear medium. As the pump propagates, the signal and idler fields are generated in phase with pump at that point in the medium. These fields then propagate to the output at their own phase velocity. Only if there is a fixed phase relationship between the pump, signal and idler, will the daughter fields generated at different points in the material constructively interfere, producing a non-zero output.

The propagation constants in Eq. 2.3 are frequency dependent and therefore the frequencies for which Eq. 2.3 is satisfied is dependent on the dispersion. For a waveguide the total dispersion has two defining contributions. Firstly, the intrinsic material dispersion and secondly the waveguide dispersion. In standard optical fibres, there is little flexibility in the waveguide contribution to the dispersion, which can only be controlled through the core size and the refractive index step. The ability to control the waveguide dispersion of PCF allows the phasematching condition to be satisfied for a set of target wavelengths through careful selection of the structural parameters of the fibre.

2.2 Photon Pair Generation in Photonic Crystal Fibres

In this section, a brief consideration of the quantum theory of FWM is made. Previously Alibart et al. [1] and others [2, 3] have presented similar calculations, in which the FWM process is studied in the interaction picture. In the framework of quantum theory, FWM can be described as the spontaneous annihilation of two pump photons, and the creation of two daughter photons, the signal and idler, spaced equally about the degenerate pump in frequency. As the photons are always generated in pairs, due to energy conservation, detection of one photon can be used to herald the presence of the remaining twin.

In this thesis, the discussion is limited to the case of co-polarised, degenerate four-wave mixing within the core of an optical fibre (specifically a photonic crystal fibre) along the z-direction. It is assumed that the fibre is single mode over the entire wavelength range of interest (which is justified for certain types of PCF [4]). Generation of the signal and idler modes is therefore made into a single well-defined spatial mode determined by the geometry of the waveguide. The z-component of the wavevectors for the pump, signal and idler fields are given by the propagation constants for the fundamental mode of the fibre as β_p, β_s, and β_i,

$$\beta_j(\omega_j) = n_{eff}(\omega_j) k_j(\omega_j), \tag{2.4}$$

where ω_j, with $j = p, s, i$ is the frequency of the field, n_{eff} is the effective index of the waveguide mode, and k_j is the free space wavevector.

The generation of the two-photon state can be found by calculating the evolution of the initial state of the signal and idler modes $|\Psi(t'=0)\rangle = |0_s, 0_i\rangle$, into the final state at time $t' = t$, $|\Psi(t'=t)\rangle$ under the action of the interaction Hamiltonian,

$$|\Psi(t)\rangle = \exp\left(\frac{1}{i\hbar} \int_0^t dt' H_I(t')\right) |\Psi(0)\rangle. \tag{2.5}$$

The FWM interaction Hamiltonian can be expressed as,

$$H_I(t') = \varepsilon_0 \chi^{(3)} \int_V dV \hat{E}_p(\vec{r}, t') \hat{E}_p(\vec{r}, t') \hat{E}_s(\vec{r}, t') \hat{E}_i(\vec{r}, t'), \tag{2.6}$$

where $\hat{E}_j$ are the quantised field operators for the pump, signal and idler respectively. Each of the field operators can be decomposed into positive and negative frequency components,

$$\hat{E}_j(\vec{r}, t') = \hat{E}_j^{(+)}(\vec{r}, t') + \hat{E}_j^{(-)}(\vec{r}, t'), \tag{2.7}$$

corresponding to the annihilation ($\hat{a}_j$) and creation ($\hat{a}_j^\dagger$) operators respectively [5], where

$$\hat{E}_j^{(+)}(\vec{r},t') = E_j(\vec{r_T}) \int_0^\infty d\omega_j A_j(\omega_j)\hat{a}_j(\omega_j)\exp(i(\beta_j(\omega_j)z - \omega_j t')) = \left(\hat{E}_j^{(-)}(\vec{r},t')\right)^\dagger, \quad (2.8)$$

with $E_j(\vec{r_T})$ is the transverse field profile of the guided mode and $A_j(\omega_j) = i\sqrt{\hbar\omega_j/2\varepsilon_0 n^2(\omega_j)}$. As the FWM process is weak, a relatively intense pump field is required. This allows the electric field operator of the pump to be described using a classical field [1, 2], where the positive frequency component is given by

$$\hat{E}_p^{(+)}(\vec{r},t') = E_p(\vec{r},t') = \tilde{\alpha}(t')E_p(\vec{r_T})\exp(i(\beta_p(\omega_p) - \gamma P_p)z), \quad (2.9)$$

where $\tilde{\alpha}(t')$ is the temporal profile of the pulse and $E_p(\vec{r_T})$ is the transverse mode profile of pump. The self-phase modulation of the pump pulse is included by the factor $\exp(-i\gamma P_p z)$, where P_p is the peak power and γ is the optical nonlinearity,

$$\gamma(\omega_p) = \frac{n_2(\omega_p)\omega_p}{cA_{eff}}, \quad (2.10)$$

where $n_2(\omega_p)$ is the Kerr-coefficient of fused silica, c is the speed of light and A_{eff} is the effective area of the mode [6]. The frequency dependence of γ can be neglected as it is slowly varying across the bandwidth of pump envelope.

The exponential in Eq. 2.5, can be expanded as,

$$|\Psi(t)\rangle \approx \left(1 + \frac{1}{i\hbar}\int_0^t dt' H_I(t')\right)|\Psi(0)\rangle, \quad (2.11)$$

where only terms $O(1)$ in the interaction Hamiltonian have been retained. Substituting Eqs. 2.7–2.9 into Eq. 2.6 and solving Eq. 2.11 yields the two-photon state which describes the signal and idler modes after the FWM interaction.,

$$|\Psi(\omega_s,\omega_i)\rangle \approx |0_s,0_i\rangle + \eta\int_0^\infty d\omega_s \int_0^\infty d\omega_i f(\omega_s,\omega_i)\hat{a}_s^\dagger(\omega_s)\hat{a}_i^\dagger(\omega_i)|0_s,0_i\rangle, \quad (2.12)$$

where $\eta = \epsilon_0\chi^{(3)}A_s(\omega_s)A_i(\omega_i)E_{p0}^2 IL/i\hbar$ is an efficiency parameter related to the strength of the non-linearity $\chi^{(3)}$, electric field amplitude of the pump E_{p0}, the overlap integral I of the mode profiles of the three fields and the length of the fibre L.

Terms $\geq O(2)$ in the expansion of Eq. 2.11 relate to the generation of multiple pairs. In general, the two-photon state is in a linear superposition of photon number states, where each ket $|n_s,n_i\rangle$ corresponds to generation of n photons into the signal and idler modes. Due to energy conservation photon number is correlated between the two modes.

Through the heralding measurement, the zero photon state will be removed and so can be eliminated from Eq. 2.12 leaving,

$$|\Psi(\omega_s,\omega_i)\rangle \approx \eta\int_0^\infty d\omega_s \int_0^\infty d\omega_i f(\omega_s,\omega_i)\hat{a}_s^\dagger(\omega_s)\hat{a}_i^\dagger(\omega_i)|0_s,0_i\rangle. \quad (2.13)$$

2.2 Photon Pair Generation in Photonic Crystal Fibres

The function $f(\omega_s, \omega_i)$ is the joint spectral amplitude (JSA) function. The joint spectral intensity (JSI) is given by $|f(\omega_s, \omega_i)|^2$ and dictates with what probability a signal photon with frequency ω_s is generated with a corresponding idler photon with frequency ω_i.

The joint spectral amplitude function is itself composed of two parts,

$$f(\omega_s, \omega_i) = \phi(\omega_s, \omega_i)\alpha(\omega_s + \omega_i), \tag{2.14}$$

$\phi(\omega_s, \omega_i)$ is the fibre phasematching function (PMF) and $\alpha(\omega_s + \omega_i)$ is the pump envelope function. Together these two functions dictate the variations in the signal and idler frequencies about their perfectly phasematched frequencies, due to the finite bandwidth of the pulsed pump envelope and the tolerable phase mismatch.

The pump envelope is commonly described with a Gaussian spectral distribution,

$$\alpha(\omega_s + \omega_i) = \exp\left(-\frac{((\omega_s - \omega_{s0}) + (\omega_i - \omega_{i0}))^2}{4\sigma_p^2}\right), \tag{2.15}$$

where ω_{p0} is the central frequency of the pump, and σ_p is the 1/e bandwidth of the pump envelope, related to the full-width half maximum $\Delta\omega_{FWHM} = 2\sigma_p\sqrt{4\ln 2}$. The phasematching function is found by integrating the phasemismatch over the length of fibre,

$$\phi(\omega_s, \omega_i) = \chi^{(3)}\int_0^L dz \exp(i\Delta\beta z), \tag{2.16}$$

with,

$$\Delta\beta = 2\beta_p(\omega_p) - \beta_s(\omega_s) - \beta_i(\omega_i) - 2\gamma P_p. \tag{2.17}$$

Assuming that $\Delta\beta$ is constant over the length of the fibre, Eq. 2.17 can be evaluated to yield,

$$\phi(\omega_s, \omega_i) = \text{sinc}\left(\frac{\Delta\beta L}{2}\right)\exp\left(i\frac{\Delta\beta L}{2}\right), \tag{2.18}$$

where L is the length of fibre, and $\Delta\beta$ is the phase mismatch between the four fields. The bandwidth of the phasematching function scales inversely with the fibre length. This is a result of the amount of phasemismatch that is acquired on propagation through the fibre. If the fibre is very long then the four fields remain phasematched over only a narrow range of frequencies due to walk-off between the pump, signal and idler pulses that is caused by dispersion. Conversely, if the fibre is very short then the total phasemismatch that is acquired remains smaller over a larger frequency range. The relationship between the pump envelope, phasematching function and joint spectrum is illustrated in Fig. 2.1 for a PCF source.

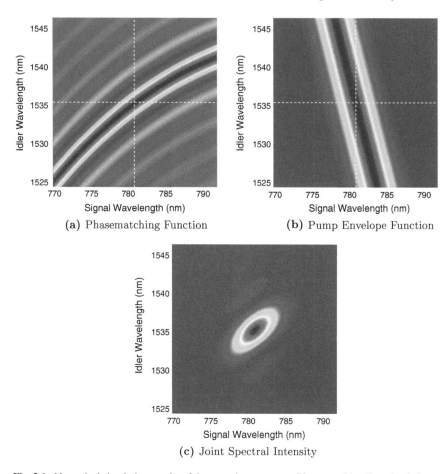

Fig. 2.1 Numerical simulation results of the two-photon state. **a** Phasematching Function **b** Pump Envelope Function **c** Joint Spectral Intensity

2.3 Spectral Engineering of the Two-Photon State

As the joint spectral amplitude is dependent on both ω_s and ω_i, the spectrum of each mode can be strongly correlated with its twin due to the energy conservation and phasematching requirements. The majority of single photon detectors are not frequency resolving, thus when one photon is detected as the herald, the twin is projected into a incoherent mixture of spectral modes. This mixture of spectral modes results in a reduction in the visibility of Hong-Ou-Mandel interference. Ultimately, for photons to be used with HOM based devices, the photon must be in a single, indistinguishable, pure state to ensure error free operation. This can only be achieved if the spectral correlations in the joint spectral amplitude can be removed.

2.3 Spectral Engineering of the Two-Photon State

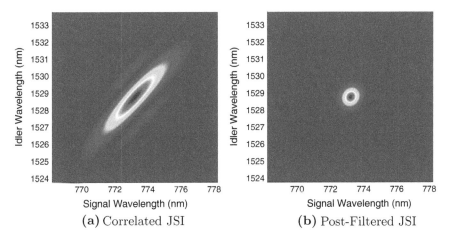

Fig. 2.2 The effect of narrowband filtering on the JSI. **a** Correlated JSI **b** Post-Filtered Decorrelated JSI

In general there are two routes to achieve this. The first is to use narrowband spectral filtering to reshape the JSA, removing the correlations and approximating a single spectral mode. However, this results in discarding a large proportion of the photon pairs generated, reducing the overall brightness or coincidence count rate of the source. In Fig. 2.2, the result of narrowband filtering on an intially correlated JSA is illustrated for two Gaussian profile spectral filters of the form:

$$f(\omega_j) = \frac{1}{\sqrt{2\pi}\sigma_j} \exp\left(\frac{-(\omega_j - \omega_{0j})^2}{2\sigma_j^2}\right). \quad (2.19)$$

This technique has been exploited to produce heralded single photon state of high purity, but at the expense of reduced count rates [7, 8].

A more suitable approach is to generate photon pairs directly in a single joint spectral mode. This requires no lossy spectral filtering; so high count rates can still be attained. It can be shown from Eq. 2.13, that a single joint spectral mode can be achieved, if and only if the joint spectral amplitude function can be factorised into two functions, one that is dependent on the frequency of the signal photon ω_s and another that is dependent on the frequency of the idler ω_i,

$$f(\omega_s, \omega_i) = g(\omega_s) \cdot h(\omega_i). \quad (2.20)$$

In this case, the detection of one photon in the pair leaves the other unaffected by the measurement. Thus on detection of the signal photon, the idler photon is heralded in a pure state. A joint spectral amplitude that displays this behaviour is described as factorable.

The correlation induced by energy conservation and imparted upon the JSA by the pump envelope function is fixed by Eq. 2.2. The orientation of the phasematching function is however determined by the dispersion of the non-linear medium, in this case the PCF. Thus by careful selection of the fibre parameters the phasematching function orientation can be selected to counteract the correlation induced by the pump envelope function. Photonic crystal fibres are an ideal pair-generation medium in this regard, due to the vast degree of flexibility of the modal dispersion that can be achieved through relatively simple control of the cladding parameters.

The effect on the orientation of the phasematching function due to the dispersion of the PCF has been studied extensively by a number of groups [9–14]. All considerations focus on achieving a factorable state by determining under what conditions the joint spectral amplitude function can be made to be separable. To begin with the propagation constants in the phasemismatch are expanded in a Taylor series about the central frequencies ω_{0j}, $j = p, s, i$, where ω_{0s} and ω_{0i} are the perfectly phasematched frequencies, and retaining terms upto to $O(1)$ corresponding to the group-velocity:

$$\beta_j(\omega_j) \approx \beta_j(\omega_{0j}) + (\omega_j - \omega_{0j}) \frac{\partial \beta_j(\omega_j)}{\partial \omega_j}\bigg|_{\omega_j=\omega_{0j}}. \tag{2.21}$$

Substituting Eq. 2.21, into Eq. 2.17, the phasemismatch becomes,

$$\Delta\beta \approx \Delta\beta^{(0)} + \varphi_s \tau_s + \varphi_i \tau_i, \tag{2.22}$$

where $\Delta\beta^{(0)} = 0$ is the phasemismatch at the perfectly phasematched wavelengths, $\varphi_j = \omega_j - \omega_{0j}$ and,

$$\tau_j = \frac{\partial \beta_p(\omega_p)}{\partial \omega_p}\bigg|_{\omega_p=\omega_{0p}} - \frac{\partial \beta_j(\omega_j)}{\partial \omega_j}\bigg|_{\omega_j=\omega_{0j}}, \tag{2.23}$$

$$= \frac{1}{\nu_p} - \frac{1}{\nu_j}, \tag{2.24}$$

where ν_j is the group-velocity of the pump, signal and idler. It can be shown that the joint spectral amplitude function is separable if,

$$\tau_s \tau_i \leq 0. \tag{2.25}$$

This constrains the potential group-velocities of the pump, signal and idler, to two classes of states. Firstly, if $\tau_s = -\tau_i$, then the group velocities must fulfil either,

$$\nu_s(\omega_s) > \nu_p(\omega_p) > \nu_i(\omega_i), \tag{2.26}$$

or,

$$\nu_i(\omega_i) > \nu_p(\omega_p) > \nu_s(\omega_s). \tag{2.27}$$

2.3 Spectral Engineering of the Two-Photon State

This is known as a symmetric group velocity matched (GVM) state, as when the bandwidths of the pump envelope and phasematching function are matched, the signal and idler photons have equal bandwidths. The second class of states are asymmetrically group velocity matched, where either $\tau_s = 0$ or $\tau_i = 0$. In this case, either the signal or idler photon propagates at the same group velocity as the pump, whilst the other walks off. Therefore a factorable state can be achieved when the pump bandwidth is sufficiently large.

The orientation of the phasematching function that corresponds to these points can be investigated by considering the angle of the phasematching function relative to the signal frequency axis,

$$\theta_{\text{pmf}} = -\arctan\left(\frac{\tau_i}{\tau_s}\right). \quad (2.28)$$

For a symmetrically GVM state $\theta_{\text{pmf}} = +45^o$, where as for asymmetrically GVM states $\theta_{\text{pmf}} = 0^o$ or 90°, corresponding to $\tau_s = 0$ and $\tau_i = 0$ respectively. On a phasematching contour (PMC) plot, such as those shown in Fig. 2.3, the asymmetric GVM states corresponds to points where the gradient of the contour is zero, here small changes in the pump wavelength produce little change in the wavelength of the non-group velocity matched photon.

Both schemes have been implemented in birefringent fibres, both commercial single mode fibres and PCF, where the birefringence is exploited to satisfy the phase- and group-velocity matching requirements to produce a state with reduced spectral correlations [15–18]. Further to this symmetric GVM sates have been attained by Soller et al. by using PCF, in which there are two neighbouring zero dispersion wavelengths (ZDW) close together. In this case, as illustrated in Fig. 2.3, the phasematching contours form closed loops, and so every possible orientation angle of the phasematching function can be selected by tuning the pump wavelength. This also serves a second purpose. By working on the outer sidebands of the phasematching contours, the phasematched wavelengths are far de-tuned from the pump into a region where the effect of noise from Raman shifted pump light is reduced. The potential factorable JSIs are illustrated in Fig. 2.3 together with the corresponding phasematching contours for a fibre displaying two neighbouring ZDWs.

In Chap. 4, the design and fabrication of a PCF exhibiting two ZDWs is presented. In the target design, photon pairs are generated at 810 and 1550 nm for the signal and idler respectively when pumped at 1064 nm by an ultrafast fibre laser. These photon pairs have been spectrally engineered in an asymmetrically GVM state to achieve a pure heralded state.

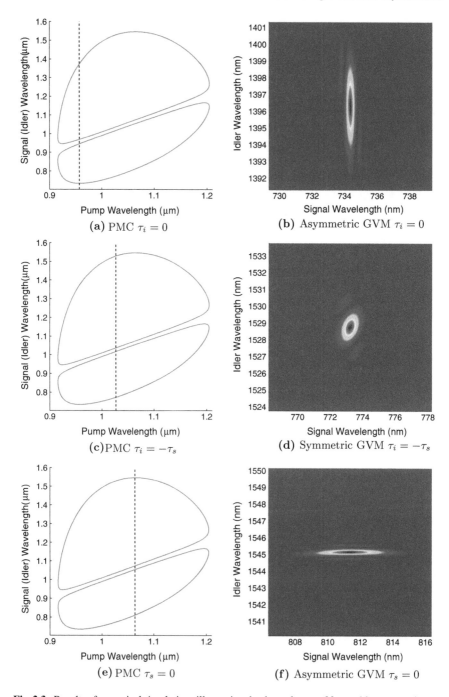

Fig. 2.3 Results of numerical simulations illustrating the three classes of factorable states. **a** Asymmetric GVM with $\tau_i = 0$, **b** Symmetric GVM $\tau_i = -\tau_s$, **c** Asymmetric GVM $\tau_s = 0$

2.4 Reduced State Spectral Purity: Schmidt Decomposition

The purity, P, of the spectrally engineered heralded state can be determined through the Schmidt mode decomposition of the two-photon joint spectral amplitude function [19, 20]. The pure two-photon state in Eq. 2.13 is an example of a bipartite state formed of two subsystems, one for the signal $\mathcal{H}_s$ and one for the idler $\mathcal{H}_i$ [21]. The complete Hilbert space of the two-photons state is,

$$\mathcal{H} = \mathcal{H}_s \otimes \mathcal{H}_i. \tag{2.29}$$

where each subsystem is described by a orthonormal basis set $|\vartheta(\omega_s)\rangle$ and $|\xi(\omega_i)\rangle$.

When a heralding measurement is made, the reduced density matrix of the heralded idler photon, $\hat{\rho}_i$, is found by tracing over $\mathcal{H}_s$ of the complete density matrix $\hat{\rho}$. The purity of the resultant heralded state is then found by taking the trace of the square of the reduced density matrix [19],

$$P = Tr\{\hat{\rho}_i^2\}. \tag{2.30}$$

If $\{|\vartheta(\omega_s)\rangle\}$ is a basis set of $\mathcal{H}_s$ and $\{|\xi(\omega_i)\rangle\}$ is a basis set of $\mathcal{H}_i$ then $\{|\vartheta(\omega_s)\rangle \otimes |\xi(\omega_i)\rangle\}$ is a basis set of $\mathcal{H}_s \otimes \mathcal{H}_i$ [21]. Therefore, the complete bipartite system can be written as a linear superposition in this basis,

$$|\Psi(\omega_s, \omega_i)\rangle = \sum_j \lambda_j^{\frac{1}{2}} |\vartheta_j(\omega_s)\rangle \otimes |\xi_j(\omega_i)\rangle, \tag{2.31}$$

where $\sum_j \lambda_j = 1$. The number of terms required to describe $|\Psi(\omega_s, \omega_i)\rangle$ in the Schmidt basis indicates how factorable the state is. The Schmidt number K is defined as,

$$K \equiv \frac{1}{\sum_j \lambda_j^2} \equiv \frac{1}{Tr\{\hat{\rho}_s^2\}} \equiv \frac{1}{Tr\{\hat{\rho}_i^2\}} \equiv \frac{1}{P}, \tag{2.32}$$

and is a measure of the number of Schmidt modes present [20]. The initial two-photon state $|\Psi(\omega_s, \omega_i)\rangle$ is said to be factorable if there is only a single pair of Schmidt modes active, in which case $\lambda_1 = 1$, $\lambda_{j \neq 1} = 0$, and $K = 1$. Therefore, by determining the Schmidt mode decomposition of the JSA we can measure the purity of the heralded idler state. This can be carried out numerically by taking the singular value decomposition of the matrix that describes the JSA [20]. We can also relate the number of active Schmidt modes and the heralded purity to the photon number statistics obtained through a measurement of the second order coherence for only a single arm.

2.5 Photon Number Statistics

The purity of the heralded state in terms of the generated photon number is a crucial measure of the source performance. The ideal single photon source would only ever generate a single photon Fock state, a state containing a definite number of photons with zero variance, with no possibility of generating more than one single photon. However, the pair generation process can yield more than one photon pair per pump pulse due to its spontaneous nature. In reality the terms that were discarded in Eq. 2.11 are important, as they describe the potential of the source to generate more than one pair per pulse. Hence the output state of the source before heralding is a linear superposition of photon number terms,

$$|\Psi\rangle = a_0|0_s, 0_i\rangle + a_1|1_s, 1_i\rangle + a_2|2_s, 2_i\rangle + \cdots, \quad (2.33)$$

where $\{a_n\}$ are the set of probability amplitude coefficients that describe the probability $|a_n|^2$ of generating n photon pairs in the interaction, with $\sum_n |a_n|^2 = 1$. For a spectrally correlated state, these coefficients are determined by a Poisson distribution, whereas for a spectrally decorrelated state they are given by a thermal distribution (where $a_n = a_1^n$ for $n \geq 1$) as discussed in Sect. 2.5.3.

When a successful heralding detection is made using a binary single photon detector (with no photon number resolving capability), the state is reduced to,

$$|\Psi\rangle = a_1|1_i\rangle + a_2|2_i\rangle + \cdots, \quad (2.34)$$

as at least one pair must have been generated. Therefore, there is the possibility that more than one photon can be heralded. Heralding multiple idler photons as a single is detrimental to the fidelity of the photon number state. This can be limited by working at low pump powers, such that the probability of generating a pair is small. This places a fundamental limit on the pair-generation rate, far from the ideal deterministic source of single photons.

2.5.1 Raman Scattering

One of the primary sources of noise in optical fibre based photon pair sources is the spontaneous Raman scattering of pump photons into the long wavelength idler mode [3, 22]. Raman scattering is another non-linear process that may occur within the fibre, caused by the coupling of the intense pump pulse to optical phonons in the material [23].

As the pulse propagates through the fibre a pump photon at ω_p is split into a Stokes photon ω_{st} and a phonon at frequency Ω, where $\omega_{st} = \omega_p - \Omega$. Through this mechanism the idler mode becomes contaminated with photons at the Stokes frequency. This increases the probability of observing a coincident detection event between the signal and idler modes, where a signal photon is detected simultaneously with a Stokes shifted photon in the idler arm. This reduces both the spectral and number purity of the heralded state and poses a significant hurdle for fibre based photon pair sources.

One route to eliminating the Stokes shifted Raman scattered light is to cool the fibres, reducing the population of the phonon modes [24–27]. However, doing this is another experimental complication, reducing the usability of the source. Instead, the dispersion of the fibre can be used to satisfy phasematching of the signal and idler modes far detuned from the pump, either by using birefringent fibre [15, 16], pumping in the normal dispersion regime [28, 29], or using a fibre with two ZDWs so that the phasematching contours for closed loops far from the pump [30].

2.5.2 Degree of Heralded Second Order Coherence

The photon number statistics of the heralded state can be determined by measuring the degree of second order coherence $g^{(2)}(0)$. For an ideal single photon Fock state $g^{(2)}(0) = 0$. The general second order coherence function, $g^{(2)}(\tau)$, is a measure of the intensity correlations of an optical mode at two times separated by a time delay τ,

$$g^{(2)}(\tau) = \frac{\langle \bar{I}(t)\bar{I}(t+\tau)\rangle}{\langle \bar{I}(t)\rangle^2}, \qquad (2.35)$$

$$= \frac{\langle E^*(t)E^*(t+\tau)E(t+\tau)E(t)\rangle}{\langle E^*(t)E(t)\rangle^2}, \qquad (2.36)$$

where $\bar{I}$ is the average intensity, E is the electric field and angled brackets $\langle \cdots \rangle$ indicate the average over a statistically significant time period much greater than the coherence time [31]. For a classical light source, Cauchy's inequality can be applied,

$$2\bar{I}(t_1)\bar{I}(t_2) \leq \bar{I}(t_1)^2 + \bar{I}(t_2)^2, \qquad (2.37)$$

to place a lower bound on the second order coherence [31]. Equation 2.37 implies that,

$$\langle \bar{I}(t)^2\rangle \geq \langle \bar{I}(t)\rangle^2. \qquad (2.38)$$

Therefore, for a classical light source at zero time delay $\tau = 0$,

$$g^{(2)}_{cl}(0) \geq 1. \qquad (2.39)$$

For any light source, the only other bound is that intensity must be positive in nature and so,

$$\infty \geq g^{(2)}_{cl}(0) \geq 1. \qquad (2.40)$$

A similar expression that is valid for quantum fields can be derived by replacing the classical electric field amplitudes in Eq. 2.36 with the relevant quantum mechanical operators [31]. If the quantum mechanical electric field operators are written in terms of the creation ($\hat{a}_j^\dagger$) and annihilation ($\hat{a}_j$) operators, an expression for the quantum degree of second order coherence, $g_{qu}^{(2)}(0)$, in terms of the photon number occupation operator can be found,

$$g_{qu}^{(2)}(0) = \frac{\langle \hat{n}(\hat{n}-1) \rangle}{\langle \hat{n} \rangle^2}, \qquad (2.41)$$

where $\hat{n} = \hat{a}^\dagger \hat{a}$. If the mode is occupied by n photons, then Eq. 2.41 evaluates to,

$$g_{qu}^{(2)}(0) = \begin{cases} 0 & \text{for } n = 0, 1 \\ (n-1)/n & \text{for } n \geq 2. \end{cases} \qquad (2.42)$$

Therefore, only a nonclassical light source capable of producing Fock states [32], where there are exactly n photons in the mode, can exhibit a $g^{(2)}(0) < 1$. Overall, the range of allowed values for $g_{qu}^{(2)}(0)$ is,

$$0 \leq g_{qu}^{(2)}(0) \leq \infty. \qquad (2.43)$$

An experimental measurement of the degree of second order coherence resulting in a $g^{(2)}(0) < 1$ is a demonstration of the generation of a non-classical state of light, where for the highest quality single photon sources we expect a value of $g^{(2)}(0) \to 0$.

Typically the degree of second order coherence of a photon pair source is measured in a Hanbury Brown–Twiss (HBT) interferometer [33]. The output mode of the source is sent to a 50:50 beamsplitter; in each of the two output ports a single photon detector is placed. These two detectors are then monitored for coincident detection events. For a perfect single photon source, where there is only one photon present in the mode in a given time bin, the two detectors will never register a simultaneous event and $g^{(2)}(0) = 0$. This measurement has become commonplace in the characterisation of heralded single photon sources and is a crucial metric by which these sources can be compared [34–36].

2.5.3 Degree of Marginal Second Order Coherence

A second form, the *marginal* second order coherence, $g_m^{(2)}(0)$, reflects the fact that only one arm of the source is measured, ignoring all heralding signals from the other arm. In doing so, the idler probability distribution is measured whilst averaging over the signal probability distribution. The goal of the phasematching scheme implemented in each photon pair source is to produce photon pairs into only two spectral-temporal modes, one for the signal and one for the idler. By performing this

2.5 Photon Number Statistics

measurement the number of spectral modes in which the idler is generated in can be experimentally determined.

For a thermal light source consisting of a single spectral mode of the electromagnetic field, the photon number statistics are described by a Bose-Einstein or thermal distribution, with a variance of

$$(\Delta n)^2 = \bar{n} + \bar{n}^2, \tag{2.44}$$

where Δn is the standard deviation of the fluctuations of the photon occupation number around the mean photon number $\bar{n}$ [31]. The degree of second order coherence calculated at zero time delay, using Eq. 2.40, for a beam of light described by Bose-Einstein statistics yields an expected a value of $g^{(2)}(0) = 2$. If the number of spectral modes is greater than one, then the effect this has on the photon number statistics can be determined from $g_m^{(2)}(0)$. Consider a collection of spectral modes of multi-mode thermal light, it can be shown that the variance of the photon number distribution is given by,

$$(\Delta n)^2 = \bar{n} + \frac{\bar{n}^2}{N_m}, \tag{2.45}$$

where N_m is the number of spectral modes present [37]. When the number of spectral modes is large, the variance and photon number statistics of the light is transformed from a thermal distribution to a Poisson distribution $(\Delta n)^2 \to \bar{n}$. If the degree of second order coherence is calculated for a beam of light described by a Poisson distribution, then the expected value is $g^2(0) = 1$ [31]. Therefore by measuring the photon number statistics of a beam of light in the temporal domain, some knowledge of the spectral content can be gained.

If we now consider the idler photons produced through pair generation and measure the photon number statistics as they exit the source we can determine the number of spectral modes into which they could be generated. If the idler photons are generated into a single spectral mode we expect $g_m^{(2)}(0) = 2$ and if there are many potential modes we expect $g_m^{(2)}(0) \to 1$. The number of modes can be extracted from the experimental value of $g_m^{(2)}(0)$ through,

$$g_m^{(2)}(0) = 1 + \frac{1}{K}, \tag{2.46}$$

where K is the Schmidt number [19, 35, 38]. For a single mode $K = 1$ and $g_m^{(2)}(0) = 2$, where as, as $K \to \infty$, $g_m^{(2)}(0) \to 1$.

References

1. O. Alibart, J. Fulconis, G.K.L. Wong, S.G. Murdoch, W.J. Wadsworth, J.G. Rarity, Photon pair generation using four-wave mixing in a microstructured fibre: theory versus experiment. New

J. Phys. **8**, 67 (2006)
2. J. Chen, X. Li, P. Kumar, Two-photon-state generation via four-wave mixing in optical fibers. Phys. Rev. A **72**, 033801 (2005)
3. Q. Lin, F. Yaman, G.P. Agrawal, Photon-pair generation in optical fibers through four-wave mixing: role of Raman scattering and pump polarization. Phys. Rev. A **75**, 023803 (2007)
4. T.A. Birks, J.C. Knight, P.S. Russell, Endlessly single-mode photonic crystal fiber. Opt. Lett. **22**, 961–963 (1997)
5. W.P. Grice, I.A. Walmsley, Spectral information and distinguishability in type-II downconversion with a broadband pump. Phys. Rev. A **56**, 1627–1634 (1997)
6. G.P. Agarwal, *Nonlinear Fiber Optics*, 4th edn., Optics and Photonics (Academic Press, San Diego, 2006)
7. T. Aichele, A. Lvovsky, S. Schiller, Optical mode characterization of single photons prepared by means of conditional measurements on a biphoton state. Eur. Phys. J. D At. Mol. Opt. Plasma Phys. **18**, 237–245 (2002)
8. P.P. Rohde, W. Mauerer, C. Silberhorn, Spectral structure and decompositions of optical states, and their applications. New J. Phys. **9**, 91 (2007)
9. K. Garay-Palmett, H.J. McGuinness, O. Cohen, J.S. Lundeen, R. Rangel-Rojo, A.B. U'ren, M.G. Raymer, C.J. McKinstrie, S. Radic, I.A. Walmsley, Photon pair-state preparation with tailored spectral properties by spontaneous four-wave mixing in photonic-crystal fiber. Opt. Express **15**, 14870–14886 (2007)
10. K. Garay-Palmett, R. Rangel-Rojo, A.B. U'Ren, Tailored photon pair preparation relying on full group velocity matching in fiber-based spontaneous four-wave mixing. J. Mod. Opt. **55**, 3121–3131 (2008)
11. O. Cohen, J.S. Lundeen, B.J. Smith, G. Puentes, P.J. Mosley, I.A. Walmsley, Tailored photon-pair generation in optical fibers. Phys. Rev. Lett. **102**, 123603 (2009)
12. K. Garay-Palmett, A.B. U'Ren, R. Rangel-Rojo, Tailored photon-pair sources based on inner-loop phasematching in fiber-based spontaneous four-wave mixing. Revista Mexicana de Física **57**, 15–22 (2011)
13. L. Cui, X. Li, N. Zhao, Minimizing the frequency correlation of photon pairs in photonic crystal fibers. New J. Phys. **14**, 123001 (2012)
14. B. Fang, O. Cohen, J.B. Moreno, V.O. Lorenz, State engineering of photon pairs produced through dual-pump spontaneous four-wave mixing. Opt. Express **21**, 2707–2717 (2013)
15. A. Clark, B. Bell, J. Fulconis, M.M. Halder, B. Cemlyn, O. Alibart, C. Xiong, W.J. Wadsworth, J.G. Rarity, Intrinsically narrowband pair photon generation in microstructured fibres. New J. Phys. **13**, 065009 (2011)
16. B.J. Smith, P. Mahou, O. Cohen, J.S. Lundeen, I.A. Walmsley, Photon pair generation in birefringent optical fibers. Opt. Express **17**, 23589–23602 (2009)
17. A.R. McMillan, J. Fulconis, M. Halder, C. Xiong, J.G. Rarity, W.J. Wadsworth, Narrowband high-fidelity all-fibre source of heralded single photons at 1570 nm. Opt. Express **17**, 6156–6165 (2009)
18. A. McMillan, Development of an all-fibre source of heralded single photons. Ph.D. thesis, University of Bath, 2011
19. A.B. U'Ren, C. Silberhorn, R. Erdmann, K. Banaszek, W.P. Grice, I.A. Walmsley, M.G. Raymer, Generation of pure-state single-photon wavepackets by conditional preparation based on spontaneous parametric downconversion, Arxiv preprint arXiv:quant-ph/0611019 (2006)
20. P.J. Mosley, J.S. Lundeen, B.J. Smith, P. Wasylczyk, A.B. U'Ren, C. Silberhorn, I.A. Walmsley, Heralded generation of ultrafast single photons in pure quantum states. Phys. Rev. Lett. **100**, 133601 (2008)
21. C.K. Law, I.A. Walmsley, J.H. Eberly, Continuous frequency entanglement: effective finite Hilbert space and entropy control. Phys. Rev. Lett. **84**, 5304–5307 (2000)
22. A. Migdall, S.G. Polyakov, J. Fan, J.C. Beinfang (eds.), Single-photon generation and detection, in *Experimental Methods in the Physical Sciences*, vol. 45 (Elsevier, 2013)
23. R. Boyd, *Non-Linear Optics*, 3rd edn. (Academic Press, Burlington, 2008)

References

24. H. Takesue, K. Inoue, 1.5-μm band quantum-correlated photon pair generation in dispersion-shifted fiber: suppression of noise photons by cooling fiber. Opt. Express **13**, 7832–7839 (2005)
25. K.F. Lee, J. Chen, C. Liang, X. Li, P.L. Voss, P. Kumar, Generation of high-purity telecom-band entangled photon pairs in dispersion-shifted fiber. Opt. Lett. **31**, 1905–1907 (2006)
26. S.D. Dyer, M.J. Stevens, B. Baek, S.W. Nam, High-efficiency, ultra low-noise all-fiber photon-pair source. Opt. Express **16**, 9966–9977 (2008)
27. S.D. Dyer, B. Baek, S.W. Nam, High-brightness, low-noise, all-fiber photon pair source. Opt. Express **17**, 10290–10297 (2009)
28. J. Rarity, J. Fulconis, J. Duligall, W. Wadsworth, P. Russell, Photonic crystal fiber source of correlated photon pairs. Opt. Express **13**, 534–544 (2005)
29. J. Fulconis, O. Alibart, W. Wadsworth, P. Russell, J. Rarity, High brightness single mode source of correlated photon pairs using a photonic crystal fiber. Opt. Express **13**, 7572–7582 (2005)
30. C. Söller, B. Brecht, P.J. Mosley, L.Y. Zang, A. Podlipensky, N.Y. Joly, P.S.J. Russell, C. Silberhorn, Bridging visible and telecom wavelengths with a single-mode broadband photon pair source. Phys. Rev. A **81**, 031801 (2010)
31. R. Loudon, *The Quantum Theory of Light*, vol. 1, 2nd edn. (Oxford University Press, Oxford, 1973)
32. L. Mandel, E. Wolf, *Optical Coherence and Quantum Optics*, 2nd edn. (Cambridge Unviersity Press, Cambridge, 1995)
33. R.H. Brown, R.Q. Twiss, Correlation between photons in two coherent beams of light. Nature **177**, 27–29 (1956)
34. X.-S. Ma, S. Zotter, J. Kofler, T. Jennewein, A. Zeilinger, Experimental generation of single photons via active multiplexing. Phys. Rev. A **83**, 043814 (2011)
35. G. Harder, V. Ansari, B. Brecht, T. Dirmeier, C. Marquardt, C. Silberhorn, An optimized photon pair source for quantum circuits. Opt. Express **21**, 13975–13985 (2013)
36. J.B. Spring, P.S. Salter, B.J. Metcalf, P.C. Humphreys, M. Moore, N. Thomas-Peter, M. Barbieri, X.-M. Jin, N.K. Langford, W.S. Kolthammer, M.J. Booth, I.A. Walmsley, On-chip low loss heralded source of pure single photons. Opt. Express **21**, 13522–13532 (2013)
37. A.M. Fox, *Quantum Optics An Introduction*, number 15 in Oxford Master Series in Physics, 1st edn. (Oxford University Press, Oxford, 2006)
38. A. Eckstein, A. Christ, P.J. Mosley, C. Silberhorn, Highly efficient single-pass source of pulsed single-mode twin beams of light. Phys. Rev. Lett. **106**, 013603 (2011)

Chapter 3
Numerical Modelling of Multiplexed Photon Pair Sources

3.1 Overview

In this Chapter, the theory and performance of two different multiplexing schemes is discussed in detail. This undertaking builds on the work of Christ and Silberhorn [1] who considered a set of spatially multiplexed PDC sources. Here we extend the consideration to include the effect that loss in imperfect components has on the probability of delivering a single photon from the scheme.

To do this, a numerical model was constructed to calculate the overall probability of successfully delivering a heralded single photon from the output of the multiplexed device. We also calculate the fidelity of the heralded state with a single photon Fock state. We have used this, together with the successful delivery probability, as a basis to compare the different source constructions. Finally, we summarise the performance enhancement achievable from each instance of the multiplexed device, by computing waiting time to deliver N_p heralded single photons, from N_p multiplexed devices.

In Sect. 3.3, a "log-tree" spatial multiplexing is considered, following this a novel temporal loop scheme is evaluated in Sect. 3.4. Recently there have been a number of manuscripts released that discuss the potential performance of multiplexed photon pair sources in spatial domain [2–4] and the temporal domain [5, 6], including my own work on which this Chapter is based. Through this body of work, we have developed a useful method by which multiplexed photon pair sources can be optimised given the unavoidable losses in the constituent components.

3.2 The Building Blocks

3.2.1 Pair Generation

We begin by considering the photon number statistics of an individual heralded single photon source based on either PDC or FWM. A schematic of such a source is shown in Fig. 3.1.

The output quantum state can be written as a superposition of photon number terms:

$$|\Psi\rangle = a_0|0_s, 0_i\rangle + a_1|1_s, 1_i\rangle + a_2|2_s, 2_i\rangle + \cdots, \quad (3.1)$$

where $\{a_n\}$ are the set of probability amplitude coefficients. If we assume that the source has been engineered such that the signal and idler photons are generated into only two spatio-temporal modes [7], and thus the two modes are only correlated in photon number, the probability amplitudes are described by Bose-Einstein (thermal) statistics [8]:

$$|a_n|^2 = p_{th}(n) = \frac{1}{(\bar{n}+1)}\left(\frac{\bar{n}}{\bar{n}+1}\right)^n, \quad (3.2)$$

where $\bar{n}$ is the mean photon number per pump pulse. For a thermal distribution of photon number, the maximum probability of generating a single pair is limited to 0.25. The vacuum component of Eq. 3.1 is eliminated through the detection of one photon in the pair to herald the presence of its twin. The remaining heralded photon is projected into a mixture of photon number states with weightings given by the set of probability amplitude coefficients $\{a_n\}$. The overall quality of the heralded single photon state will be limited by contributions from higher photon number terms as well as the reintroduction of the vacuum component through loss in any components

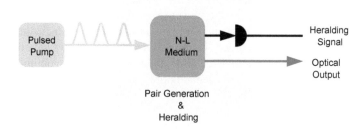

Fig. 3.1 A heralded single photon source. A pulsed laser is used to pump a non-linear optical medium, the daughter photon pairs generated by either FWM or PDC are split in wavelength. The signal photon (*black arrow*) is sent to a detector where a successful detection event results in a heralding signal. The corresponding heralded idler photon (*red arrow*) is routed to the optical output of source

3.2 The Building Blocks

between point the of generation and the output. We define the signal-to-noise ratio (SNR) as the relative contribution of single to multi-photon terms in the heralded state,

$$\text{SNR} = \frac{P(1)}{\sum\limits_{n=2}^{\infty} P(n)} = \frac{P(1)}{1 - P(0) - P(1)}, \quad (3.3)$$

where $P(n)$ is the probability per pulse of delivering n photons from the output arm. To allow a meaningful comparison between different multiplexed source configurations, i.e. detector and switch efficiencies and number of sources, we determine the value of $\bar{n}$ that corresponds to the same SNR in all source configurations.

3.2.2 Detection

Following pair generation the signal and idler photons are split, and the signal photon is sent to a detector to act as the herald for the remaining idler photon. We will consider the effect of using three different heralding detectors, binary, PNR and pseudo-PNR. The heralding measurement performed by the detector is described by a set of positive operator value measure (POVM) elements. The general form of such a POVM for a detection outcome of 'n' is given by,

$$\hat{\Pi}(n) = \sum_{N=n}^{\infty} p_{det}(n|N)|N\rangle\langle N|, \quad (3.4)$$

where $p_{det}(n|N)$ is the conditional probability of the detector recording result 'n' from 'N' photons at the input [9]. The form of this conditional probability is dependent on the type and mode of operation of the detector.

For a binary detector there are only two possible detection outcomes. Either "click" or "no-click", therefore the set of POVM operators contains only two elements, corresponding to the different detection outcomes:

$$p_{det}(\text{"click"}|N) = \left[1 - (1 - \eta_d)^N\right], \quad (3.5)$$

$$p_{det}(\text{"no-click"}|N) = (1 - \eta_d)^N, \quad (3.6)$$

where η_d is the lumped efficiency of the detector which can include all losses in the channel leading to the detector.

A PNR detector can be approximated by splitting the detector input into a number of distinct spatial or temporal modes, with each mode monitored by a binary detector. The splitting is typically accomplished by using a network of free-space beamsplitters or, for convenience, 50:50 fibre couplers. Provided that all routes have the same splitting ratio, the probability of a photon incident at the input occupying any one of the final modes is proportional to the inverse of the number of modes. One of the fundamental limitations of such a pseudo-PNR detector is that its operation is non-deterministic, with the potential for more than one photon to end up in a single mode, leading to the detector under-counting. For a detector with some degree of photon number resolving capability the set of POVM elements extends over the possible values of photon number, up to the limit of the number of detection modes present in the device. A typical pseudo-PNR constructed by O'Sullivan et al. [10] and Coldenstrodt-Ronge et al. [11], where the detector input is split over M modes, each monitored by a binary detector can be represented by the set of POVM elements for measuring n photons with conditional detection probabilities:

$$p_{det}(n|N) = \binom{M}{n} \sum_{j=0}^{n} (-1)^j \binom{n}{j} \left((1-\eta_d) + \frac{\eta_d(n-j)}{M} \right)^N. \quad (3.7)$$

By letting the number of modes extend to infinity the set of POVM elements of a true PNR detector with conditional detection probabilities of the form,

$$p_{det}(n|N) = \binom{N}{n} \eta_d^n (1-\eta_d)^{N-n}, \quad (3.8)$$

can be recovered.

We can now use the relevant POVM element to determine the form of the heralded idler state following the heralding detection of n_s signal photons, by projecting the chosen POVM element onto the single-mode pair state of Eq. 3.1 and tracing over the detected signal photon:

$$\hat{\rho}_i(n_s) = \frac{Tr_s\{\hat{\Pi}(n_s)|\Psi\rangle\langle\Psi|\}}{\langle\Psi|\hat{\Pi}(n_s)|\Psi\rangle}, \quad (3.9)$$

This then defines an ensemble of possible reduced density matrices for the heralded idler state, depending on the detection result in the signal arm. From this we can calculate the SNR of the heralded state by selecting the reduced density matrix corresponding to a successful heralding event and calculating the ratio of the single-photon to multi-photon terms. For a binary detector this becomes:

$$\text{SNR} = \frac{\langle 1_i|\hat{\rho}_i(\text{"click"})|1_i\rangle}{\sum_{q=2}^{\infty} \langle q_i|\hat{\rho}_i(\text{"click"})|q_i\rangle}, \quad (3.10)$$

where as for a PNR detector it becomes:

$$\text{SNR} = \frac{\langle 1_i|\hat{\rho}_i(n_s=1)|1_i\rangle}{\sum_{q=2}^{\infty} \langle q_i|\hat{\rho}_i(n_s=1)|q_i\rangle}. \quad (3.11)$$

3.2 The Building Blocks

Here the SNR was set to a value of 100 to allow direct and easy comparisons between different source configurations. This corresponds to a 1% error probability in line with the level of fault tolerance of the computational scheme developed by Knill [12]. However, not all applications require such a high level of tolerance, for example quantum key distribution is routinely quoted as only requiring an SNR of 10 for a usable system [13, 14].

3.2.3 Metrics and Measurements

In order to make direct comparisons between different detector configurations we set the SNR to a constant value and determine the corresponding mean photon number per pulse. At the corresponding mean photon number per pulse we can then calculate the fidelity, F, of the heralded state to a pure single photon state,

$$F = \frac{\langle 1|\hat{\rho}_i(n_s = 1)|1\rangle}{Tr\{\hat{\rho}_i(n_s = 1)\}}, \tag{3.12}$$

and the probability of successfully delivering a single photon at this mean photon number.

We identify a successful outcome of the source as a single heralding (PNR) or "click" (binary) event leading to a single heralded photon at the output of the source. To find the overall probability of successfully delivering a single photon from the source, p(success), we first determine the probability of a successful heralding event, p(heralding), and then multiply by the fidelity of the heralded idler state. In doing so we are selecting the reduced density matrix corresponding to a successful heralding event and multiplying by the probability that there is a successful heralding event. For a PNR detector, a successful heralding event constitutes the detection of one photon in the signal arm, thus using Eqs. 3.1, 3.4, and 3.8, p(heralding) $= p(n_s = 1) = \langle \Psi|\hat{\Pi}(1)|\Psi\rangle$ and we have:

$$p(\text{success}) = p(n_s = 1)\frac{\langle 1|\hat{\rho}_i(n_s = 1)|1\rangle}{Tr\{\hat{\rho}_i(n_s = 1)\}}, \tag{3.13}$$

whereas for a binary detector p(heralding) $= p(\text{"click"}) = \langle \Psi|\hat{\Pi}(\text{"click"})|\Psi\rangle$ and:

$$p(\text{success}) = p(\text{"click"})\frac{\langle 1|\hat{\rho}_i(\text{"click"})|1\rangle}{Tr\{\hat{\rho}_i(\text{"click"})\}}. \tag{3.14}$$

From these success probabilities we can now calculate the mean waiting time, t_{wait}, to deliver N_p independent photons from N_p independent sources,

$$t_{wait} = \frac{1}{p(\text{success})^{N_p} \cdot R_p}. \tag{3.15}$$

Calculating these metrics at a fixed SNR and determining the corresponding mean photon number per pump pulse, $\bar{n}$, will allows us to make direct comparisons between different configurations of the sources. In all that follows, we define the multiplexing depth of the complete device, as the number of switching stages between a single source and the output in the spatial scheme, and the number of time bins multiplexed over in the temporal domain.

3.3 Spatial Multiplexing

Using the building blocks developed in the preceding section we can now begin to construct the multiplexed device. We will initially only consider a 2×1 multiplexing scheme as shown schematically in Fig. 3.2, from here we can then cascade pairs of sources up to some requested number. The output of each individual source, each with its own heralding detector, are coupled to long lengths of optical fibre to delay the heralded photon for enough time for the switching electronics to set the state of the switch. These fibre delays feed directly to the input of the 2×1 optical switch. The state of the switch is controlled by the heralding detectors. When the heralding detector from source 1 fires the switch is set to allow the channel 1 input through to the output and similarly for source 2.

As the switch has some finite switching bandwidth or rise time, this limits the rate at which we can pump the sources. Using a laser with a higher repetition rate than the switch bandwidth will result in the possibility of missing a heralded photon. The

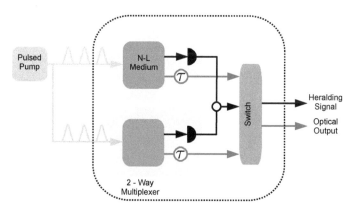

Fig. 3.2 A 2×1 spatially multiplexed heralded single photon source. Two individual sources are pumped simultaneously by the same pulsed laser. The generated signal photons in each source are routed to two detectors. The idler photons are coupled into long lengths of optical fibre to delay the photon by the time required to set the state of the switch, before being fed into a 2×1 optical switch. The state of the switch is controlled by the heralding signals from the two sources, e.g. If the heralding detector of source 2 fires the switch routes the idler photon from source 2 to the common output and vice versa for source 1

3.3 Spatial Multiplexing

switch cannot be set to the correct state in enough time to encompass the arriving heralded photon. As commercially available switches currently operate at a repetition rate of around 1 MHz we will limit the repetition rate of the pump laser to also 1 MHz. Therefore, overall, when one detector produces a successful heralding signal, the switch is set to route the corresponding heralded state to the common output, and the channel from the other source remains closed.

To find the overall probability of successfully delivering a single photon from the multiplexed device we must determine the two parts of the product in Eq. 3.13. Firstly, the probability that at least one of the two sources produces a heralding signal. Secondly, the fidelity of the heralded state given loss in the delay lines and the switch.

To find the probability that at least one of the sources produces a heralding signal, we begin by defining a heralding probability vector, whose elements correspond to a particular POVM element. For example, if the heralding detector is a PNR detector we have:

$$\vec{p_H} = \begin{pmatrix} \langle \Psi | \hat{\Pi}(0) | \Psi \rangle \\ \langle \Psi | \hat{\Pi}(1) | \Psi \rangle \\ \langle \Psi | \hat{\Pi}(2) | \Psi \rangle \\ \vdots \\ \langle \Psi | \hat{\Pi}(n) | \Psi \rangle \end{pmatrix}, \quad (3.16)$$

where the first row corresponds to zero photons detected, the second row corresponds to 1 photon detected etc. The equivalent heralding probability vector for a source utilising binary detection contains only 2 rows, corresponding to "click" and "no-click".

From $\vec{p_H}$ we can determine all the possible combinations of detection events distributed across the two sources by taking the Kronecker product of the two heralding probability vectors (one for each source):

$$p_H^{(2)} = \vec{p_H}^T \otimes \vec{p_H} = \begin{pmatrix} P_{00} & P_{10} & P_{20} & \cdots & P_{n0} \\ P_{01} & P_{11} & P_{21} & \cdots & P_{n1} \\ P_{02} & P_{12} & P_{22} & \cdots & P_{n2} \\ \vdots & \vdots & \vdots & \ddots & \vdots \\ P_{0m} & P_{1m} & P_{2m} & \cdots & P_{nm} \end{pmatrix}, \quad (3.17)$$

where P_{nm} corresponds to source 1 heralding 'n' photons and source 2 heralding 'm' photons. By summing elements of this matrix we can recover a new heralding probability vector that describes the multiplexed device. All those elements highlighted in blue correspond to the probability of at least one of the sources signalling a successful heralding event, thereby the switch selects the corresponding source and the density matrix $\hat{\rho}_i(n_s = 1)$ is selected from the ensemble described earlier.

The full heralding probability vector can be re-established by applying some selection rules, all those remaining elements whose index contains a zero are summed to give the new probability of observing no heralding detection. The other elements, in the bottom right hand corner, can then be summed up based on the lowest value of the two indices, i.e. P_{32} is included in the probability of heralding two photons etc.

Following a successful heralding event, we now need to take into account the effect of the loss in the components between source and switch and the switch itself, and how this affects the density matrix of the heralded idler state. To do this, the delay line is modelled as having an efficiency of η_τ and the switch efficiency as η_s. The total loss between the point of generation to the output of the complete multiplexed system can be treated as an unbalanced beamsplitter with a transmission coefficient, $\mathcal{T} = \sqrt{\eta}$ and reflectivity coefficient of $\mathcal{R} = \sqrt{1-\eta}$ where the concatenated loss $\eta = \eta_\tau \cdot \eta_s^{N_s}$ and $N_s = \log_2(N)$ is the number of switches required for N sources. Here we have assumed that all the components are identical, in principle it is possible to extend this treatment to non-identical sources and routings. The transmitted state, which becomes the heralded output including loss, is found by applying the general beamsplitter transformation for an input containing n_i photons:

$$|n_i^A, 0_i^B\rangle = \sum_{p=0}^{n} \binom{n}{p}^{\frac{1}{2}} \eta^{\frac{p}{2}} (1-\eta)^{\frac{n-p}{2}} |p_i^C, (n-p)_i^D\rangle, \qquad (3.18)$$

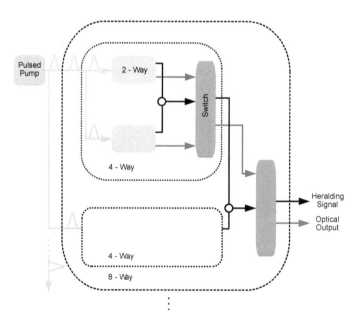

Fig. 3.3 Schematic of a cascaded array of spatially multiplexed heralded single photon sources

3.3 Spatial Multiplexing

where A, B and C, D label the input and output ports of the beamsplitter respectively. The reduced density matrix for the heralded idler photon including loss is found by tracing over the loss mode D, $|(n-p)_i^D\rangle$ and then renormalising. Finally, the probability of successfully delivering a single photon can be found using Eq. 3.13.

The above treatment can be extended to an arbitrary number of sources by cascading them in pairs as shown in Fig. 3.3. The heralding probability vector can be determined by applying Eq. 3.17 recursively, using the reconstructed heralding vector from the previous pair of sources to solve for the two-pairs of two sources etc. The reduced density matrix of the heralded idler photon is found in the same manner as before, where now we take into account an extra set of switches and delays.

3.3.1 Simulation Results and Discussion

3.3.1.1 Signal-to-Noise

To provide a meaningful comparison between different source architectures comprising different detectors and losses we choose to evaluate the key metrics at a fixed SNR. The mean photon number that yields this level of SNR is then used to evaluate

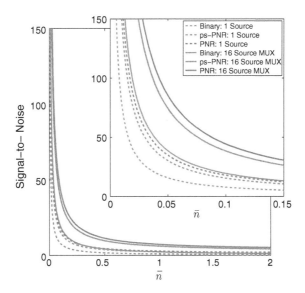

Fig. 3.4 Comparison of achievable SNR for individual sources (*dashed*) and 16-way multiplexed systems (*solid*) for three different heralding detectors: Binary (*blue*), pseudo-PNR (*green*) and PNR (*red*). The performance at low average photon number is shown inset. At high mean photon numbers the SNR is dominated by multi-photon contributions, as the mean photon number is lowered the SNR increases due to the reduction in multi-pair events. All calculations carried out a detector efficiency $\eta_d = 70\%$, switch efficiency $\eta_s = 80\%$ and delay $\eta_\tau = 99\%$

p(success) and the fidelity. Firstly, the SNR is calculated for a range of mean photon number in the interval $\bar{n} = 0$ to $\bar{n} = 2$.

Figure 3.4 compares the results of the simulations for individual sources and 16-way multiplexed sources with binary, pseudo-PNR and PNR detectors with efficiency η_d. At high mean photon numbers the overall heralded idler state is dominated by large contributions due to multi-photon generation resulting in a low SNR for all the sources. As the mean photon number is reduced the SNR increases for all sources as the noise is reduced.

Comparing between the different detector types shows that, as expected, a PNR detector offers greater SNR for fixed mean photon number due to its ability to discriminate heralding events from multiple photon pairs. At these efficiency levels, the binary detector exhibits the poorest SNR with the 8-bin pseudo-PNR outperforming it. Following the multiplexing there is an overall increase in the SNR as the overall probability of successfully heralding a single photon from any of the sources is increased. We choose to compare systems at a fixed SNR at a value of 100 and adjust the mean photon number accordingly.

The single pair per pulse generation probability, given by a thermal distribution, and the overall probability of success is shown in Fig. 3.5. The reduction in noise that is achieved by moving from a binary detector to one with PNR capabilities, allows each source to operate at a higher mean photon number whilst maintaining the same SNR. Similarly, an increase in the single pair generation probability is seen on moving to a multiplexed device, again due to the ability to pump a higher average power, whilst maintaining the target SNR. These effects, coupled with the gains made through the multiplexing protocol, result in a significant increase in the overall probability of successfully delivering a single photon from the system.

The effect of detector efficiency on the achievable SNR is explored in Fig. 3.6. Here, the mean photon number is fixed at the value that yields an SNR of 100 at $\eta_d = 0.7$. The detection efficiency for a 16-way multiplexed system is then scanned

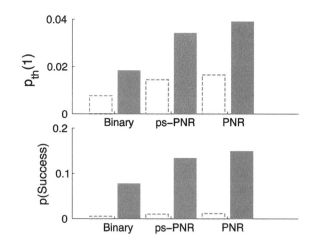

Fig. 3.5 *Top* Single-pair generation probabilities at mean photon numbers corresponding to a SNR of 100, *Bottom* probability of successfully delivering a single photon from the output of the scheme at a SNR of 100. Individual source (*dashed*), 16-Way multiplexed system (*solid*) for three different detector mechanisms: Binary (*blue*), pseudo-PNR (*green*), PNR (*red*)

3.3 Spatial Multiplexing

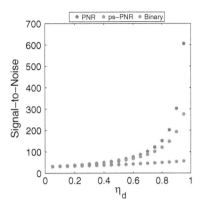

Fig. 3.6 Behaviour of Signal-to-Noise with detection efficiency at a mean photon number corresponding to a SNR of 100 at $\eta_d = 70\%$

and the resultant SNR calculated. As one would expect, the SNR of the system increases with increasing detector efficiency. The rapid increase in SNR observed for the PNR detector is again due to its ability to discriminate and reject those heralding detection events corresponding to more than one photon. Conversely, for a binary detector, where no information regarding the number of photons is known, the curve is much flatter. For low mean photon numbers the performance of a 8-bin pseudo-PNR detector is very similar to that of a true PNR detector up to detection efficiencies in the region of 80%. At the selected SNR value of 100 and the accompanying value of mean photon number, Fig. 3.7 shows the overall probability of delivering n_i photons from an individual source and a 16-way multiplexed device.

In order to minimise the contributions of higher-order photon-number components when using binary detectors, one is forced to operate at low mean photon numbers, yielding a source in which a single pump pulse will most likely produce nothing at all; only on a small fraction of pulses will the detector "click", as seen in Fig. 3.7a. By incorporating a further 15 sources into one 16-way multiplexed system the probability of successfully delivering a heralded single photon is increased, see Fig. 3.7b. By employing a PNR detector the overall probability of successfully delivering a single photon is in the first instance greater than that for a source with binary detectors, Fig. 3.7c, and the beneficial effect of multiplexing 16 sources is clearly seen in Fig. 3.7d. In both cases, this overall increase comes without a corresponding increase in the noise from higher-order photon-number components.

3.3.1.2 Effect of PNR Detector Efficiency

In order to explore the effect of detector efficiency on overall multiplexed source performance, we restrict the discussion to the use of only PNR detectors in the heralding arms of each source. Again, we determine the value of the mean photon number at a fixed SNR of 100, but now fixing the delay line and switch efficiencies at $\eta_\tau = 99\%$ and $\eta_s = 80\%$ respectively, whilst allowing the detector efficiency to vary.

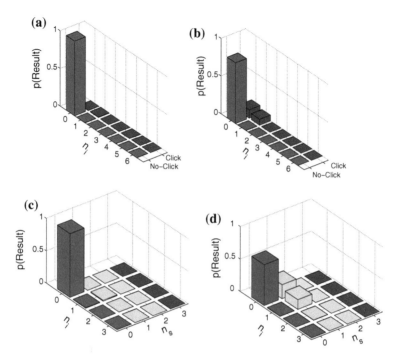

Fig. 3.7 Probability of delivering n photons given a particular heralding event, at $\bar{n}$ corresponding to an SNR of 100. Horizontal axes denote the resultant idler state and heralding detection results, vertical axis denotes the overall probability. Simulations carried out with efficiencies $\eta_d = 70\%$, $\eta_\tau = 99\%$ and $\eta_s = 80\%$. **a** Single source binary detector, **b** 16-way multiplexed system, binary detector, **c** Single source PNR detector, **d** 16-Way multiplexed system, PNR detector

Figure 3.8 shows how the heralding probability, fidelity and overall probability of success vary with mean photon number for 3 different detector efficiencies, $\eta_d = 30$, 70 and 100%.

In the first column, the probability of a successful heralding event, p(heralding) is explored. As more sources are included in the multiplexed system p(heralding) increases. At low mean photon numbers, the initial rate at which p(heralding) increases, rises with increasing detector efficiency. For a perfect PNR detector the maximum value of p(heralding) will always occur at $\bar{n} = 1$ as this is the peak of the thermal distribution for a single-pair generation event. As the detector efficiency drops, the peak value of p(heralding) shifts to higher $\bar{n}$ as the detector begins to miss single-pair generation events and provides more spurious "successful" heralding events resulting from multi-pair generation events.

In the second column, the fidelity of the heralded idler state is calculated according to Eq. 3.12. A perfect PNR detector ($\eta_d = 100\%$) can always distinguish between single- and multi-pair generation events, as a result the fidelity of the resultant heralded idler state becomes independent of $\bar{n}$. The fidelity is instead fixed at a constant value that is determined by the concatenated loss between point of generation and the

3.3 Spatial Multiplexing

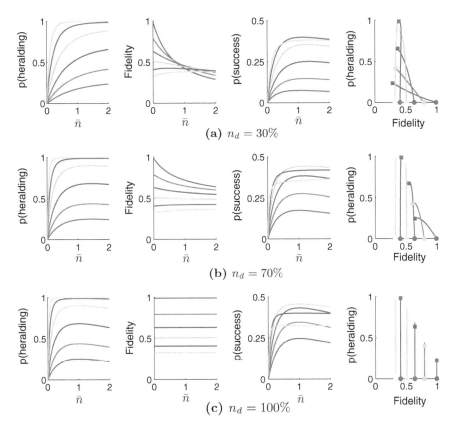

Fig. 3.8 Multiplexed source performance for varying detector efficiency with fixed switch efficiency $\eta_d = 80\%$. **a** $\eta_d = 30\%$, **b** $\eta_d = 70\%$, **c** $\eta_d = 100\%$. *First column* probability of a successful heralding event, p(heralding), as a function of mean photon number $\bar{n}$; *second column* fidelity of the heralded output state as a function of $\bar{n}$; *third column* probability of successfully delivering a single photon, p(success) = p(heralding) × fidelity, as a function of $\bar{n}$; *fourth column* trade-off between p(heralding) and fidelity between $\bar{n} = 0$ (*circles*) and $\bar{n} = 2$ (*squares*). Colours indicate number of individual sources in each multiplexed system: 1 (*blue*), 2 (*green*), 4 (*red*), 8 (*cyan*), 16 (*purple*) and 32 (*yellow*)

common output. Through the optical losses of the constituent source components, the vacuum component of the state vector is reintroduced to the heralded state leading to a reduction in the overall fidelity. Therefore by increasing the multiplexing depth, incorporating more lossy switches, actually leads to a reduction in the realisable fidelity of the output state. For detectors with less than unit efficiency, as $\bar{n} \to 0$ the contributions due to multi-pair events which are normally the largest contributor to the reduction of the fidelity become negligible, and the fidelity tends towards the same value as in the case of perfect detection.

For large multiplexing depths with imperfect detectors and switches (Fig. 3.8a, b), the fidelity may counter-intuitively increase over a small range of increasing mean photon number. In this instance, a multi-pair generation event may be incorrectly labelled as a successful heralding event, as all but one of the signal photons are lost leading to only one of the photons being detected. Similarly, the loss of the delay-line and the switching network can lead to the loss of all but one of the idler photons, resulting in the probabilistic conversion of a multi-pair generation event into a successful outcome.

The overall probability of delivering a single photon from the multiplexed system is shown in the third column of Fig. 3.8, and results from the combination of the heralding probability and the fidelity in columns 1 and 2 respectively. For a source using perfect detectors the optimal mean photon number resulting in the highest probability of delivering a heralded single photon state is found at $\bar{n} = 1$, as this is the peak of the thermal distribution. Moving away from perfect detectors, this value shifts to slightly higher mean photon numbers reflecting the behaviour of the heralding probability. Furthermore, there is no guarantee that increasing the multiplexing depth will translate into an increase in the probability of success. This effect is especially notable for detectors with high detection efficiency and at mean photon numbers in excess of 0.1. Here, the overall probability of success is limited by the loss in the switching network, which will increase with multiplexing depth. Ultimately, p(success) becomes limited by the fidelity.

The fourth and final column, displays the achievable heralding probability for a given fidelity of a heralded state, and allows direct comparison to the work of Christ and Silberhorn [1]. The ideal deterministic source of single photons lies in the top right corner. Here, one and only one photon will be emitted from the source on every pump pulse from the laser. For sources constructed from imperfect components there is an inherent trade-off between the heralded fidelity and the probability of delivery. For a fixed detector efficiency, significant gains in the heralding probability can be made through multiplexing but at the expense of a slightly poorer fidelity due to switch loss.

3.3.1.3 Effect of Switch Efficiency

A similar analysis of switch efficiency on multiplexed source performance can also be made, simulation results are shown in Fig. 3.9. Again, only PNR detectors are considered and the mean photon number is calculated at a SNR of 100. The efficiencies of the detector and delay line are fixed at $\eta_d = 70\%$ and $\eta_\tau = 99\%$.

As the detector efficiency has now been fixed at $\eta_d = 70\%$ the heralding probability displayed in the first column does not vary with switch efficiency. Similarly, the fidelity of the heralded state for a single source (second column, blue) remains unchanged with switch efficiency, as there are no switches used. For a perfect switch with unit transmission $\eta_s = 100\%$ the fidelity of the heralded state becomes independent of switch efficiency and the curves for all multiplexed systems overlap. As with the case of detector efficiency, as $\bar{n} \to 0$ the fidelity tends to the total transmission of the switching network and similar loss effects can transform multi-pair generation events into single photon outcomes in the presence of high switch loss.

3.3 Spatial Multiplexing

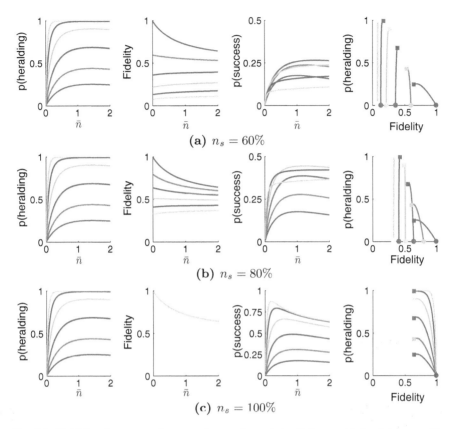

Fig. 3.9 Multiplexed source performance for varying switch efficiency with fixed detector efficiency $\eta_d = 70\%$ and delay line efficiency $\eta_\tau = 99\%$. *Top row* switch efficiency $\eta_s = 60\%$; middle row $\eta_s = 80\%$; *bottom row* $\eta_s = 100\%$. Column and colour format as in Fig. 3.8

The overall effect of switch efficiency on the probability of delivering a heralded single photon at the output is shown in the third column. For high levels of switch loss ($\eta_s \approx 60\%$), increasing the multiplexing depth by incorporating more switches (and sources) is only beneficial up to two switches (four sources); after this the total concatenated loss in the switching network limits the achievable fidelity. However, for switches with high switch efficiencies, increasing the multiplexing depth always yields an increase in the probability of successfully delivering a single photon. Using a switch with as little insertion loss as possible is crucial to high performance operation.

Finally in column four, the heralding probability is plotted against the fidelity of the heralded state. It is clear to see that at low switch efficiency, although by multiplexing many sources higher heralding probabilities can be achieved this is very much at the expense of the fidelity of the final state. However, at high switch efficiency and a large multiplexing depth the multiplexing system approaches a deterministic single photon source.

3.3.1.4 Optimisation of Multiplexed Systems

To fully realise the potential of source multiplexing and to achieve near deterministic operation, the efficiencies of all the components must be as near unity as possible. For the ideal case of a perfect PNR detector with unit detection efficiency, the signal to noise is infinite as the system can always discriminate between single photon and multi-photon generation events. This, coupled with perfect switching corresponds to the case considered by Christ and Silberhorn [1]. As seen in Fig. 3.9 it is always beneficial to increase the multiplexing depth with perfect switches. By then pumping at the peak of the thermal photon number distribution corresponding to $\bar{n} = 1$ and $p_{th}(1) = 0.25$ an overall single photon delivery probability in excess of 0.99 is achieved.

The final columns of Figs. 3.8 and 3.9 allow a comparison to be made between the results presented here and those of Christ and Silberhorn. By including the effect of loss and inefficient components, the overall source performance is degraded but remains consistent with their previous work. In addition, Figs. 3.8 and 3.9 deliver the perhaps counter-intuitive message that it is not always beneficial to multiplex as many sources as possible or operate at a mean photon number per pulse that yields the highest single pair generation probability. Using realistic components where some degree of loss is inevitable requires the multiplexed system to be optimised in such a way that the mean photon number is set in accordance with the loss of the components used. The methods and results described above give a robust method by which this optimisation can be done. For a SNR of 100, with current commercially available detector technologies the mean photon number is constrained to be low, this is the regime in which the majority of contemporary sources operate; in this case it is nearly always beneficial to have the highest multiplexing depth possible.

The results of the above analysis can be summarised by using the per-pulse probability of delivering a heralded single photon state and calculating the waiting time (Eq. 3.15) to deliver N_p photons simultaneously from N_p independent systems, presented in Fig. 3.10. This also allows s direct comparison between different multiplexed systems utilising different detector technologies and efficiencies. For the benchmark of a single source, we assume that a laser repetition rate of 76 MHz is used in line with a Ti:Sapphire laser system, where as for multiplexed systems where the repetition rate is limited by the switch bandwidth a laser repetition rate of 1 MHz is used.

For small numbers of independent photons, the waiting time may be shorter for a single source than for a multiplexed system. This is a result of the increased repetition rate at which the non-multiplexed source can be pumped. However, for delivering large numbers of photons simultaneously, multiplexed systems far surpass a single source. For example, the average waiting time for 8 heralded single photons delivered simultaneously is reduced from approximately 400 years for 8 individual sources

3.3 Spatial Multiplexing

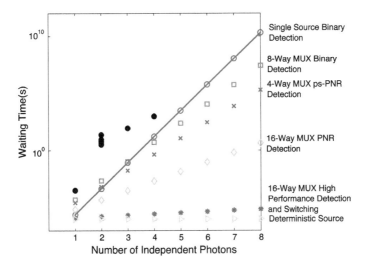

Fig. 3.10 Comparison of waiting times to deliver a number of independent heralded single photons for different systems at SNR = 100 and $\eta_d = 70\%$ unless stated otherwise. Single sources with binary detection pumped at $R_p = 80$ MHz (*blue*). Multiplexed sources running at $R_p = 1$ MHz: 8-way with binary detection (*green*), 4-way with pseudo-PNR detection (*red*), 16-way with PNR detection (*cyan*), 16-way with PNR detection for a potential future high-performance system with $\eta_d = 98\%$, $\eta_\tau = 99\%$ and $\eta_s = 95\%$ (*magenta*) and deterministic system with lossless detection, routing and switching (*yellow*). *Black circles* indicate approximate experimentally measured waiting times for N-independent photons for non-multiplexed sources, Ref. [7, 15–19]

to a few minutes for 8 realistic multiplexed sources. Looking to the future where high performance PNR detectors with high detection efficiencies ($\eta_d > 98\%$) and ultra-low loss switches ($\eta_s = 95\%$) may become available, a realistic future system consisting of 16-sources could approach deterministic operation.

3.4 Temporal Loop Multiplexing

In the preceding section active spatial multiplexing of heralded single photon sources was shown to offer many benefits, chiefly that of an increase in the delivery probability of a heralded single photon state at the output. However, this is not without a cost in the form of the large number of detectors, switches and non-linear media, of which every single one must produce the same state in all possible degrees of freedom. This is not an easy task to achieve. Following this, in order to deliver N_p individual heralded single photons requires N_p multiplexed systems with each system comprised of a potential 17 sources (for an ideal deterministic system). In total, that equates to approximately $17 \times N_p$ non-linear media. Whilst bringing significant performance benefits, spatial multiplexing has a large resource overhead.

As an alternative, one can combine the output from different temporal modes of the same source, known as temporal multiplexing. Now, the overhead cost is not in additional physical components but in the number of potential time bins/repetition rate reduction. In this section, a temporal multiplexing scheme utilising only a single source and optical switch together with a optical fibre storage loop is presented. The performance of the scheme is evaluated in a similar manner as for the spatial multiplexing cases above.

Figure 3.11 shows the proposed scheme. As in the case of spatial multiplexing a non-linear medium is pumped by a pulsed laser. The daughter photon pair is split in wavelength, again the signal photon is sent to a detector to act as the heralding signal for the remaining idler photon. In this discussion only a detector with PNR capabilities is considered, but the analysis can be simply extended to other detector types. The heralded idler photon passes into a length of fibre to delay it by enough time for the switch to be set. Following this, a 2 × 2 optical switch is used to either route the photon to the common output or to feed it into a fibre storage loop, where the loop consists of a length that delays the photon by exactly one period of laser pulse train. In doing so, photons stored in the loop are made to overlap with the next pulse from the laser.

Overall the multiplexing scheme can be described as follows. Consider a train of m laser pulses, labelled from $t = 1$ to $t = m$ in order of arrival at the non-linear medium. If a pair is generated as the first pulse propagates through the medium the heralding detector fires and switch is set to the crossed state to route the heralded idler photon into the loop. On the next pulse of the laser, if there is a second heralding event, the switch is set to the crossed state simultaneously routing the new photon into the storage loop and ejecting the previously stored photon into a rejected time bin. If there is no heralding event, the switch remains closed and the photon in the storage loop completes another pass through the switch and fibre. This is then repeated up to and including the mth pulse.

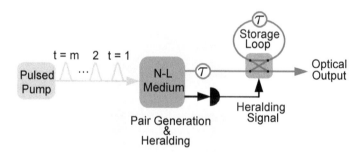

Fig. 3.11 Schematic of the temporal loop multiplexing scheme. A pulsed laser pumps a non-linear pair-generation medium. Daughter photon pairs are split in wavelength; the signal photon is routed to a detector to acts as the heralding and switch logic signal. The idler photon is routed through a long length of fibre to give enough time for the switch to be set. If a heralding signal is produced in the signal arm, the 2 × 2 switch is set to route the corresponding idler photon into the storage loop whose delay is exactly one period of the laser train. See text for full details of the scheme

3.4 Temporal Loop Multiplexing

On the mth pulse, if there is no successful heralding event then the switch is set to route the stored photon into the optical output. However, if there is a successful heralding event on the final pulse the switch remains closed allowing the newly generated photon to propagate to the output. In this way the photon amplitude that has incurred the minimum possible amount of loss is always routed to the output. However, the photon delivered from the scheme after m pulses may have passed through the switch any where from $t = 1$ to $t = m$ times, and so accrued different amounts of loss.

For low values of t a photon must make many more passes through the loop and switch to be used after the mth pulse, therefore its overall contribution to the probability of success is smaller compared to photons generated on later pulses where $t \rightarrow m$. The insertion loss of the switch and loop therefore have a large effect on the probability of success. The magnitude of this effect can be determined by carrying out a similar set of simulations, as for the spatial scheme, where we will use the same basic building blocks. As before, we assume that the source has been engineered to generated signal and idler photons into only two spatio-temporal modes so that the signal and idler modes are only correlated in photon number, and so the probability amplitude coefficients are described by thermal statistics, Eq. 3.2.

A very simple result can be found by only considering the first non-zero term in the state vector of Eq. 3.1. The overall probability of successfully delivering a single photon by multiplexing over m pulses, is the probability that all of the previous pulses don't fail to store a photon, or

$$p(\text{success}) = 1 - \prod_{t=1}^{m}\left(1 - p_{th}(1)\eta_d \eta_L^t\right), \tag{3.19}$$

where η_d and η_L are the detector efficiency and lumped efficiency of switch and storage loop. As we have neglected higher order terms in Eqs. 3.1, 3.19 is only valid at low values of $\bar{n}$.

To more faithfully describe the scheme at higher mean photon numbers, we can apply the same methods developed for spatial multiplexing to determine the reduced density matrix, $\hat{\rho}_i(n_s, t)$, that describes the heralded idler state for a given heralding detection of n_s photons on the tth pulse and delayed until the mth bin. The effect of loss is again modelled as a beamsplitter with a transmission coefficient given by the lumped efficiency of t passes through the switch.

For the tth pulse there exists a set of density matrices, $\{\hat{\rho}_i(n_s, t)\}$ with $\text{Tr}\{\hat{\rho}_i(n_s, t)\} = 1$, for a given heralding detection event of n_s photons. By using a PNR detector, all detection events where $n_s \neq 1$ can be ignored. The probability of successfully delivering a heralded single photon from the tth pulse is found by multiplying the probability of a successful heralding event with the overlap between a single photon Fock state and the density matrix for the tth pulse:

$$p(\text{success}, t) = p(n_s = 1)\frac{\langle 1|\hat{\rho}_i(n_s = 1, t)|1\rangle}{\text{Tr}\{\hat{\rho}_i(n_s = 1, t)\}}. \tag{3.20}$$

The overall probability of successfully delivering a heralded single photon from the scheme, where the pulse train has been divided into different temporal bins each containing m pulses is found through the product,

$$p(\text{success}, m) = 1 - \prod_{t=1}^{m} (1 - p(\text{success}, t)). \quad (3.21)$$

To calculate the SNR we will also need to determine the contribution due to noise, defined as the probability of a successful heralding event followed by the delivery of an idler state containing more than one photon:

$$p(\text{noise}, m) = 1 - \prod_{t=1}^{m} \left(1 - \sum_{n=2}^{\infty} p(n_s = 1) \frac{\langle n|\hat{\rho}_i(n_s = 1, t)|n\rangle}{\text{Tr}\{\hat{\rho}_i(n_s = 1, t)\}}\right). \quad (3.22)$$

When $\hat{\rho}_i$ is properly normalised such that $\text{Tr}\{\hat{\rho}_i\} = 1$, this is reduced to:

$$p(\text{noise}, m) = 1 - \prod_{t=1}^{m} \left(\sum_{n=0}^{1} p(n_s = 1)\langle n|\hat{\rho}_i(n_s = 1, t)|n\rangle\right). \quad (3.23)$$

The accuracy of both $p(\text{success}, m)$ and $p(\text{noise}, m)$ will depend on the value of n at which the calculation is truncated due to the effect of the normalisation of $\hat{\rho}_i$. Therefore the accuracy of the results of the simulation must be balanced with the computational running time. Finally, we define the signal-to-noise as,

$$\text{SNR} = \frac{p(\text{success}, m)}{p(\text{noise}, m)}, \quad (3.24)$$

again we determine the mean photon number that corresponds to a fixed level of SNR. This allows us to make comparisons between different configurations of the system and also to compare back to the spatial multiplexing schemes.

3.4.1 Simulation Results and Discussion

3.4.1.1 Signal-to-Noise

We once again choose to determine the behaviour of the SNR with varying mean photon number to identify the value of $\bar{n}$ at a SNR of 100. Figure 3.12 shows the probability of success and the achievable SNR of the system at $\eta_d = 70\%$ and $\eta_L = 80\%$.

3.4 Temporal Loop Multiplexing

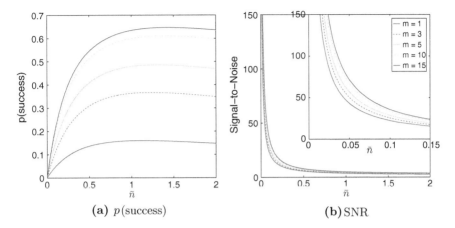

(a) p(success)

(b) SNR

Fig. 3.12 **a** Variation of $p(\text{success})$ with $\bar{n}$ for multiplexing depths of $m = 1$ (*blue*), 3 (*green*), 5 (*red*), 10 (*teal*) and 15 (*purple*). **b** The corresponding signal-to-noise of the output state, inset: SNR at low mean photon numbers. All simulations carried out with a detector efficiency of $\eta_d = 70\%$ and lumped loop efficiency $\eta_L = 80\%$

The behaviour of both the probability of success and the SNR closely resembles that of the spatial multiplexing scheme, with some exceptions. For spatial multiplexing, working at large multiplexing depths with an imperfect switch can lead to a reduction in $p(\text{success})$ compared to lower multiplexing depths. For temporal multiplexing this is no longer the case. As the number of pulses is increased, the contribution from those early pulses becomes negligible, but still acts in a positive manner, and those later pulses carry the majority of the benefit as they comprise the paths with the least amount of loss. As a result of this $p(\text{success})$ will tend towards some value at a fixed $\bar{n}$ as the multiplexing depth is increased. However, by operating at large multiplexing depths, the overall rate at which a heralded photon can be delivered becomes small as the effective repetition rate of the laser is reduced. As before, we are constrained to work in the regime of small $\bar{n}$ in order to limit contributions from multi-pair generation events and maintain a high SNR. Overall, by multiplexing, an increase in the probability of delivering a heralded single photon from output is observed at a fixed SNR.

Computing the reduced density matrix of the heralded idler state for a large number of terms is quite intensive, whereas the analytical result of Eq. 3.19 is simple, therefore it is interesting to know over what range of $\bar{n}$ it remains valid. It is also important to know at which point the state vector can be safely truncated, without neglecting too many higher order events. We can investigate the truncation effects by comparing the results of Eq. 3.19 and also the full density matrix method of Eq. 3.21. Figure 3.13 shows the difference between $p(\text{success})$ for different truncation points as a function of $\bar{n}$.

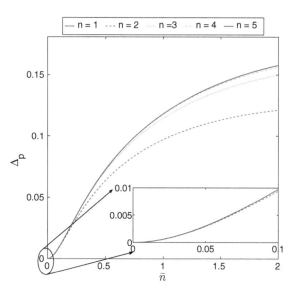

Fig. 3.13 The difference (Δ_p) between p(success) calculated using the simple analytical model of Eq. 3.19 and the full numerical calculation using the density matrix. Colours indicate the different truncation points used in the state vector, $|1_s, 1_i\rangle$ (*blue*), $|2_s, 2_i\rangle$ (*green*), $|3_s, 3_i\rangle$ (*red*), $|4_s, 4_i\rangle$ (*teal*) and $|5_s, 5_i\rangle$ (*purple*). Beyond the $n = 5$ the truncation error becomes sufficiently small as can be neglected

At high mean photon numbers there is a significant difference between the analytical model and the density matrix method. This is expected, as at these high values of $\bar{n}$, multi-pair generation events play a significant role. At low mean photon numbers there is much better agreement between the methods. As the truncation points is moved to higher terms, the difference in the value of p(success) gets larger compared to the analytical model. However the relative increase at each new truncation point reduces each time, as p(success) asymptotes to its value calculated using an infinite number of terms. Truncating at $|5_s, 5_i\rangle$ compared to $|4_s, 4_i\rangle$ shows little difference over the complete range of $\bar{n}$, especially in the region of low mean photon numbers which is of primary interest. All results presented here are calculated with the truncation point set up to and including $|5_s, 5_i\rangle$.

The behaviour of p(success) with increasing multiplexing depth is shown in Fig. 3.14 for two different regimes. Firstly, at a fixed SNR of 100 and secondly, at fixed $\bar{n}$ shown as blue circles and green squares respectively. Results are shown for a detector efficiency of $\eta_d = 70\%$ and lumped loop efficiency $\eta_L = 80\%$. In both regimes there is an increase in the probability of successfully delivering a heralded single photon. For fixed SNR, as the multiplexing depth increases, the mean photon number can be increased. As a result of this, the observed increase in p(success) is greater than that of a source held at a fixed mean photon number. Also shown in Fig. 3.14 are points corresponding to a spatially multiplexed source utilising the same multiplexing depth, detector and switch efficiencies, for fixed SNR (red triangles) and fixed $\bar{n}$ (teal triangles). It can be seen that the proposed temporal loop scheme yields a similar increase p(success) at approximately the same multiplexing depth and SNR, but with a large reduction in the amount of resources required to physically implement it.

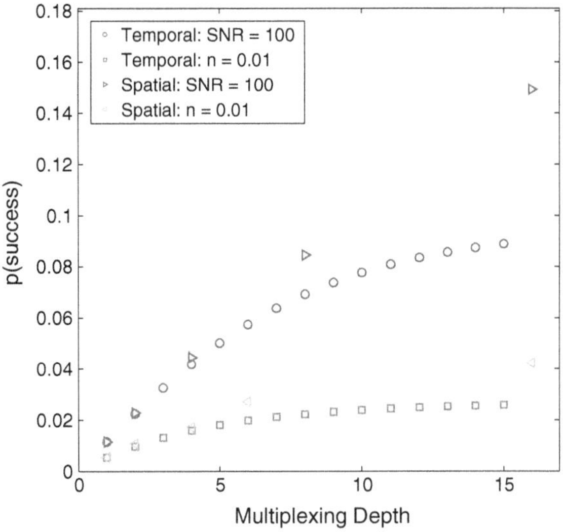

3.4 Temporal Loop Multiplexing

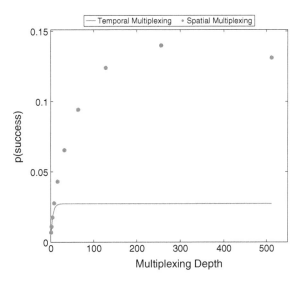

Fig. 3.14 The increase of $p(\text{success})$ with increasing multiplexing depth at a fixed $\bar{n} = 0.01$ (*green squares*) and fixed SNR $= 100$ (*blue circles*). *Triangles* show the comparison with a spatial multiplexing scheme for fixed $\bar{n}$ (*teal*) and fixed SNR $= 100$ (*Red*). Detector efficiency $\eta_d = 70\%$ and lumped loop efficiency $\eta_L = 80\%$

Fig. 3.15 The increase of $p(\text{success})$ with increasing multiplexing depth at a fixed $\bar{n} = 0.01$ for temporal and spatial multiplexing shown in *blue* and *green* respectively. Detector efficiency $\eta_d = 70\%$ and lumped loop efficiency $\eta_L = 80\%$

As previously stated, the contribution to the overall probability of success from early pulses reduces as the multiplexing depth increases due to the presence of imperfect components. This manifests itself in the saturation of $p(\text{success})$ with multiplexing depth. This can be clearly seen in Fig. 3.15 in blue, for those points at a fixed mean photon number. However, this is not true for a spatially multiplexed source as shown in green in Fig. 3.15. As the multiplexing depth is increased further, the probability of success rapidly increases, before eventually turning over and decreasing. This effect can be attributed to the increase in loss between the point of generation and the output, as more switches must be incorporated to combine more sources together.

Therefore, for a fixed level of switch loss the multiplexing depth of a spatially multiplexed source must be carefully selected so as to ensure that more loss is not inadvertently included in the device. If this can be achieved, then spatial multiplexing offers a greater benefit in source performance compared to temporal multiplexing but at the cost of more pair generation media and switches. At small values of $\bar{n} \simeq 0.01$, as the multiplexing depth is increased, $p(\text{success})$ tends to a constant value given by

$$\lim_{m \to \infty} \{p(\text{success}, m)\} = \left(\frac{\eta_L}{1 - \eta_L}\right) p(\text{heralding}). \qquad (3.25)$$

This is particularly clear in Fig. 3.15 where up to 15 sources the two schemes are comparable in performance. But for large number of sources the temporal multiplexing scheme saturates and asymptotes according to Eq. 3.25, where as the performance of a spatially multiplexing scheme continues to increase, before reducing due to the extra loss incurred. However, for a realistic device fabricating many hundreds of identical sources may be infeasible. In which case for small values of $\bar{n}$ and multiplexing depths upto 16 temporal multiplexing yields a useful performance enhancement comparable to that of a spatial multiplexing scheme.

3.4.1.2 Resource Scaling

Temporal loop multiplexing could yield an increase in $p(\text{success})$ that is commensurate with spatial multiplexing schemes of similar depths, but at a far smaller component cost. For example, at a fixed SNR of 100, a total storage loop and switch loss of $\eta_L = 80\%$ yields a performance improvement factor of approximately 6, for a multiplexing depth of $m = 8$ pulses. To construct this requires only a single photon pair source, one heralding detector and one switch. In comparison, to achieve a similar increase for a spatially multiplexed source requires vastly more resources, namely: 8 photon pair sources (which must be identical), 8 detectors and 8 delay lines and 7 switches. Building such a network would be extremely expensive and complex, as all the sources and delay lines must be matched so that the all of the heralded photons are in an identical pure state. The resource scaling for spatially and temporally multiplexed sources are compared in Table 3.1.

It is important to note that for a multiplexing depth of m, the effective repetition rate of the source is reduced to R_p/m. As a result of this the number of time bins in a which a photon could be delivered has reduced by a factor of m. Because of this, the number of heralded single photon states delivered in one second from this scheme may be surpassed by a non-multiplexed photon pair source pumped by a high-repetition rate laser such as a Ti:Sapph or Vertical External Cavity Emitting Laser (VECSEL) [20]. However, if our end goal is the simultaneous delivery of several independent photons from independent sources, this scheme presents a significant improvement over current source architectures.

3.4 Temporal Loop Multiplexing

Table 3.1 Multiplexing scheme performance comparison. Improvement calculated for multiplexing depth $m = 8$, relative to single source with a heralding detector of efficiency $\eta_d = 0.7$ and switch efficiency $\eta_L = 0.8$

Scheme	Sources	Heralding detectors	Switches	Rep. rate	SNR	$\bar{n}$	Improvement
Temporal	1	1	1	R_p/m	100	–	6.06
Spatial	2^m	2^m	2^m-1	R_p	100	–	7.40
Temporal	1	1	1	R_p/m	–	0.01	3.32
Spatial	2^m	2^m	2^m-1	R_p	–	0.01	4.03

3.4.1.3 Optimisation of Temporal Loop Scheme

As with spatial multiplexing, it is clear that in order to fully extract the potential of multiplexed single photon sources, the loss from all the components should be as minimised. For temporal loop multiplexing, where the switch is used for multiple passes, switch loss is absolutely critical. Considering the inevitable imperfections of currently-available components, the analysis presented here demonstrates that significant gains in source performance can be achieved through this scheme.

Figure 3.16a shows the variation of p(success) with detector efficiency at a fixed switch efficiency of $\eta_L = 80\%$. At low detection efficiency, the detector is unable to distinguish between single and multi-pair generation events. The source must be operated at very low $\bar{n}$ in order to maintain a useful SNR, this severely inhibits p(success). As the detector efficiency increases, p(success) rises rapidly as higher mean photon numbers can be accessed whilst maintaining a usable SNR. As $\eta_d \to 1$, the source can be operated at the peak of the thermal photon number distribution. At this level, p(success) is only limited by loss in the switch and delay line.

Figure 3.16b shows the variation of p(success) with switch efficiency at a fixed detector efficiency of $\eta_d = 70\%$. Provided that the switch and delay line have an efficiency greater than 50% then multiplexing is beneficial. However, at this level of loss, it is not worthwhile to multiplex over a large number of pulses. For high switch efficiency, p(success) becomes limited by the heralding detector as we must operate at a mean photon number yielding a sufficiently high SNR. But, the source can then be multiplexed over a large number of pulses to raise the overall probability of success.

We can investigate the performance of a future realistic low-loss system, with optimised detector and switch efficiencies of $\eta_d = 98\%$ and $\eta_L = 95\%$ respectively, shown in Fig. 3.17a. There is a clear increase in p(success) with increasing multiplexing depth, over the entire range of $\bar{n}$. With a depth of 15 pulses, a delivery probability of over 90% is achievable. At these levels of loss, the depth could be increased further yielding a pseudo-deterministic source. Finally, Fig. 3.17b shows the reduction in waiting time to deliver N_p independent photons from N_p independent sources. Due to the reduction in repetition rate from the multiplexed device compared to a single 76 MHz source, the temporal loop scheme is marginally outperformed at low numbers of requested photons. However, as the number of requested photons is increased, even a source constructed using currently available technology vastly outperforms a single source. There is clear potential for a near deterministic device, with optimised detector and switch efficiencies, constructed from relatively few components.

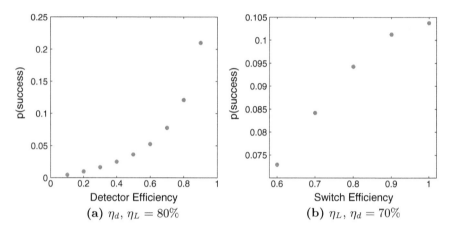

Fig. 3.16 a Effect of detector efficiency, η_d, on p(success) for a fixed switch efficiency ($\eta_L = 80\%$) at fixed SNR = 100. b Effect of switch efficiency, η_L, for a fixed detector efficiency ($\eta_d = 70\%$) at fixed SNR = 100

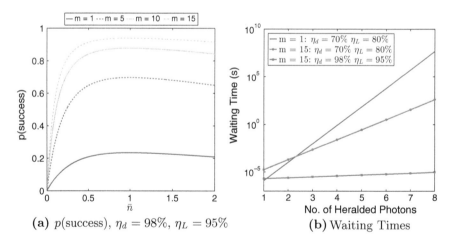

Fig. 3.17 a p(success) scaling with multiplexing depth for a future optimised device with $\eta_d = 98\%$ and $\eta_L = 80\%$, for multiplexing depths $m = 1$ (*blue-solid*), 5 (*green-dashed*), 10 (*red-dot*) and 15 (*teal-dot-dash*). b Waiting time to deliver N_p independent photons from N_p independent sources. A single 76 MHz source (*blue*), temporal loop scheme with a depth of $m = 15$ with $\eta_d = 70\%$ and $\eta_L = 80\%$ (*green*) and a future optimised device with $\eta_d = 98\%$, $\eta_L = 95\%$ and $m = 15$ (*red*)

3.5 Conclusion

In this chapter, the potential for a near-deterministic single photon source through active multiplexing in either the spatial or temporal domain was discussed. We have developed a robust manner in which a multiplexed single photon source may be optimised whilst taking into account the inevitable loss in the constituent components as well as different heralding detectors. This optimisation will be crucial in order to fully realise the potential of a multiplexed source.

By taking account of imperfect components, a meaningful comparison between sources constructed of different detectors and different efficiencies is difficult. We chose one key metric, the signal-to-noise, to be kept constant to facilitate a comparison, requiring the use of numerical techniques to evaluate source performance. We have shown that the effect of optical losses on the heralded idler state limits the extent to which the multiplexing depth can be increased for a spatial source. At high levels of loss and high mean photon numbers it is no longer beneficial to multiplex as many sources as possible. Conversely, for a temporal loop scheme there is no such reduction in performance by over multiplexing. Instead the benefit accrued from photons generated from the first pulses in the bin becomes negligible and so the increase in performance will saturate rather than reduce as the multiplexing depth is increased.

We have seen that temporal loop multiplexing is a strong contender for future multiplexed single photon sources, providing a similar performance increase as spatial multiplexing but without the monetary and resource costs. Looking to the future, it is most likely that a near-deterministic single-photon source will utilise a combination of temporal and spatial multiplexing. Several individual sources could be constructed with temporal loops attached to the output that then feed into a log-tree switch network. Fortunately, the logical signals required to set all the switches can be easily derived from the output of the heralding detectors. The above treatment of the individual multiplexing schemes can be easily extended to describe a composite system.

Finally, we have shown that source multiplexing combined with realistic improvements in detector and switch technology over the coming years is a promising candidate for supplying high-quality heralded single photons for future quantum enhanced technologies.

References

1. A. Christ, C. Silberhorn, Limits on the deterministic creation of pure single-photon states using parametric down-conversion. Phys. Rev. A **85**, 023829 (2012)
2. P. Adam, M. Mechler, I. Santa, M. Koniorczyk, Optimization of periodic single-photon sources. Phys. Rev. A **90**, 053834 (2014)
3. R.J.A. Francis-Jones, P.J. Mosley, Exploring the limits of multiplexed photon-pair sources for the preparation of pure single-photon states, ArXiv e-prints (2014)
4. D. Bonneau, G.J. Mendoza, J.L. O'Brien, M.G. Thompson, Effect of loss on multiplexed single-photon sources, ArXiv e-prints (2014)

5. P.P. Rohde, L.G. Helt, M.J. Steel, A. Gilchrist, Multiplexed single-photon state preparation using a fibre-loop architecture, ArXiv e-prints (2015)
6. R.J.A. Francis-Jones, P.J. Mosley, Temporal loop multiplexing: a resource efficient scheme for multiplexed photon-pair sources, ArXiv e-prints (2015)
7. P.J. Mosley, J.S. Lundeen, B.J. Smith, P. Wasylczyk, A.B. U'Ren, C. Silberhorn, I.A. Walmsley, Heralded generation of ultrafast single photons in pure quantum states. Phys. Rev. Lett. **100**, 133601 (2008)
8. R. Loudon, *The Quantum Theory of Light*, vol. 1, 2nd edn. (Oxford University Press, Oxford, 1973)
9. A. Feito, J.S. Lundeen, H. Coldenstrodt-Ronge, J. Eisert, M.B. Plenio, I.A. Walmsley, Measuring measurement: theory and practice. New J. Phys. **11**, 093038 (2009)
10. M.N. O'Sullivan, K.W.C. Chan, V. Lakshminarayanan, R.W. Boyd, Conditional preparation of states containing a definite number of photons. Phys. Rev. A **77**, 023804 (2008)
11. H.B. Coldenstrodt-Ronge, J.S. Lundeen, K.L. Pregnell, A. Feito, B.J. Smith, W. Mauerer, C. Silberhorn, J. Eisert, M.B. Plenio, I.A. Walmsley, A proposed testbed for detector tomography. J. Mod. Opt. **56**, 432–441 (2009)
12. E. Knill, Quantum computing with realistically noisy devices. Nature **434**, 39–44 (2005)
13. A. Migdall, S.G. Polyakov, J. Fan, J.C. Beinfang (eds.), Single-photon generation and detection, in *Experimental Methods in the Physical Sciences*, vol. 45 (Elsevier, 2013)
14. A. McMillan, Development of an all-fibre source of heralded single photons. Ph.D. thesis, University of Bath, 2011
15. B. Smith, P. Mosley, J. Lundeen, I. Walmsley, Heralded generation of two-photon NOON states for precision quantum metrology, in *Conference on Lasers and Electro-Optics, 2008 and 2008 Conference on Quantum Electronics and Laser Science, CLEO/QELS 2008* (2008), pp. 1–2
16. C.-Y. Lu, X.-Q. Zhou, O. Guhne, W.-B. Gao, J. Zhang, Z.-S. Yuan, A. Goebel, T. Yang, J.-W. Pan, Experimental entanglement of six photons in graph states. Nat. Phys. **3**, 91–95 (2007)
17. R. Kaltenbaek, B. Blauensteiner, M. Żukowski, M. Aspelmeyer, A. Zeilinger, Experimental interference of independent photons. Phys. Rev. Lett. **96**, 240502 (2006)
18. F.W. Sun, B.H. Liu, Y.F. Huang, Z.Y. Ou, G.C. Guo, Observation of the four-photon de Broglie wavelength by state-projection measurement. Phys. Rev. A **74**, 033812 (2006)
19. J.-W. Pan, Z.-B. Chen, C.-Y. Lu, H. Weinfurter, A. Zeilinger, M. Żukowski, Multiphoton entanglement and interferometry. Rev. Mod. Phys. **84**, 777–838 (2012)
20. O.J. Morris, R.J. Francis-Jones, K.G. Wilcox, A.C. Tropper, P.J. Mosley, Photon-pair generation in photonic crystal fibre with a 1.5GHz modelocked VECSEL. Opt. Commun. **327**, 39–44 (2014). special Issue on Nonlinear Quantum Photonics

Chapter 4
Design, Fabrication, and Characterisation of PCFs for Photon Pair Generation

4.1 Overview

In this chapter the design, fabrication and characterisation of a PCF for photon pair generation is described. The PCF structure was carefully designed to generate the long wavelength photon (idler) near 1550 nm, with the corresponding signal photon near 800 nm. The 1550 nm wavelength region is of particular importance in the telecommunications industry as it lies in the low loss window of silica optical fibres. In addition to this, due to the long standing history of the telecommunications industry there are many existing fibre components that can be easily integrated with low loss at a low cost, such as wavelength division multiplexers, polarisers, optical switches and delays. By targeting the signal photon at 800 nm, places it firmly within the high detection efficiency of relatively inexpensive silicon single photon avalanche photodiodes. High detection efficiency here is key, as the signal photons will act as the herald for the idler photons.

Further to targeting the idler photon at 1550 nm, particular care was taken to design a PCF structure that minimises the spectral correlations of the two-photon state produced from the FWM, see Sect. 2.3. This extra step is required to herald idler photons in pure states which are necessary for many quantum technologies applications. Section 4.2 describes the design and fabrication of the PCF to be used as the pair-generating medium. The degree of spectral correlation in the two-photon state generated by the fibre was characterised by measuring the joint spectral intensity distribution using stimulated emission tomography, see Sect. 4.3.

Photonic crystal fibres are an excellent medium with which to generate photon pairs due to the ability to control the dispersion through relatively simple changes in the structural parameters of the fibre. This yields a large degree of flexibility in the choice of pump, signal and idler wavelengths that can be used. However, in principal it is easier to design a PCF to target specific wavelengths, and then use the limited control of the pump laser to affect subtle changes in the signal and idler wavelengths.

The target wavelengths of the source were chosen to be 810 and 1550 nm for the signal and idler photons respectively when pumped at 1064 nm.

4.2 Photonic Crystal Fibre Design and Fabrication

In step- or graded-index optical fibres the refractive index contrast between the core and cladding is achieved by doping either the fused silica with another element, such as Ge in the core. In a photonic crystal fibre the refractive index contrast is instead achieved by incorporating channels of air in the cladding that run along the entire length of fibre, see Fig. 4.1. Therefore, the refractive index of the cladding lies somewhere between that of silica and air. By varying the size and separation of the air holes we can gain some degree of control over it. This in turn allows us to control the effective index of the guided mode, and engineer the waveguide contribution to the fibre dispersion. This gives us the ability to tailor the dispersion of the fibre to achieve the phasematching necessary to produce photons at our desired wavelengths with no spectral correlations.

As discussed in Sect. 2.3, in order to produce photon pairs which exhibit no spectral correlations in the two-photon state, the group velocities of the signal or idler should be matched to that of the pump. This can be achieved by producing a PCF that has two zero dispersion wavelengths fairly close together. The resultant phasematching contours form two closed loops one for the signal and one for the idler photon. The task of designing the fibre then falls to finding what PCF cladding structures yield

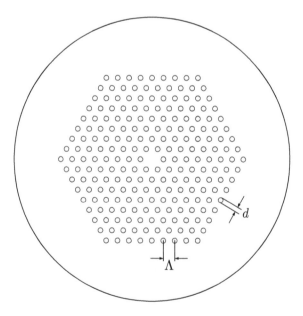

Fig. 4.1 A schematic of the structure of a photonic crystal fibre. The fused silica background is shown in *white*. The *triangular* lattice of air holes, with a diameter d and separated by the pitch Λ are shown in *light blue*. Through control of $\{d, d/\Lambda\}$ the waveguide contribution to the dispersion can be tailored

4.2 Photonic Crystal Fibre Design and Fabrication

ZDWs, with the correct gradient of phasematching contours to give group velocity matched solutions of the signal and idler at 810 and 1550 nm.

4.2.1 Simulation and Design

A simple numerical model of the PCF cladding structure and the resulting modal effective index profile was developed in Matlab using the results of Saitoh and Koshiba [1]. From the modal effective index, the propagation constant of the mode can be easily found,

$$\beta(\omega) = n_{eff}(\omega) k_0(\omega), \qquad (4.1)$$

where $k_0(\omega)$ is the free space wavevector at angular frequency ω. By taking the derivative of β with respect to ω, the group-velocity ($\beta_1 = 1/\nu_g$), and group-velocity dispersion (β_2) can be calculated,

$$\beta_1 = \frac{1}{c}\left(n_{eff}(\omega) + \omega \frac{dn_{eff}}{d\omega}\right), \qquad (4.2)$$

$$\beta_2 = \frac{1}{c}\left(2\frac{dn_{eff}}{d\omega} + \omega \frac{d^2 n_{eff}}{d\omega^2}\right), \qquad (4.3)$$

where c is the speed of light in vacuum. From these values we can then calculate the wavelengths generated through the FWM interaction in a PCF with a specific cladding structure. This is done by calculating the phasemismatch that accrues between the four-fields during propagation along the fibre,

$$\Delta\beta = 2\beta_p(\omega_p) - \beta_s(\omega_s) - \beta_i(\omega_i) - 2\gamma P_p, \qquad (4.4)$$

only those frequencies that remain phasematched ($\Delta\beta = 0$) will be coherently generated. Equation 4.4 is then evaluated for sets of three frequencies $\{\omega_p, \omega_s, \omega_i\}$ whilst maintaining energy conservation $2\omega_p = \omega_s + \omega_i$ and a contour plot of points with $\Delta\beta = 0$ can be plotted. The points of intersection of this contour with a line at the pump frequency yields the phasematched frequencies of the FWM.

Next, the joint spectral amplitude must be evaluated to ensure that at the phasematched frequencies there are no spectral correlations in the two-photon state. This is achieved by multiplying the phasematching function of the fibre $\phi(\omega_s, \omega_i)$, using the phasemismatch calculated above, with the envelope function of the pump spectrum $\alpha(\omega_s + \omega_i)$,

$$\text{JSA} = \alpha(\omega_s + \omega_i) \times \phi(\omega_s, \omega_i), \qquad (4.5)$$

for details of these functions see Sect. 2.2. The number of spectral modes can be found by applying the technique of Schmidt mode decomposition (see Sect. 2.4); from this the projected purity of the heralded idler state can be calculated. A targeted search

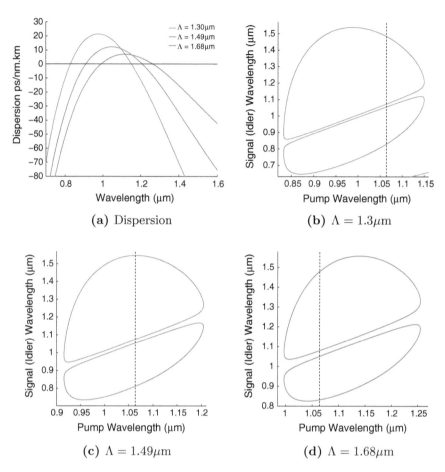

Fig. 4.2 Results of numerical simulations of the group-velocity dispersion (**a**), and resultant four-wave mixing phasematching contours (**b–d**), of a photonic crystal fibre with a fixed hole size of $d = 0.641\,\mu$m and varying pitch: **b** $\Lambda = 1.3\,\mu$m, **c** $\Lambda = 1.49\,\mu$m and **d** $\Lambda = 1.68\,\mu$m. Reducing the pitch causes the zero-dispersion wavelengths to shift to shorter wavelengths, allowing us to control the gradient of the phasematching contours at the pump wavelength and therefore the orientation of the phasematching function

of the $\{\Lambda, d/\Lambda\}$ parameter space was made, looking for phasematched wavelengths at 810 and 1550 nm, whilst maximising the purity of the heralded state.

Through a combination of small changes in $\{\Lambda, d/\Lambda\}$ a dispersion profile that satisfies our requirements was found. Figures 4.2a and 4.3a show the effect of changes in Λ and d/Λ respectively. By moving the ZDWs through changing the pitch, the gradient of the phasematching contour at the pump wavelength can be controlled. This allows a portion of the contour which corresponds to a GVM solution, such as, where the gradient of the contour is either $0°$ or $+45°$. Then by scaling d/Λ the desired phasematched wavelengths can be achieved.

4.2 Photonic Crystal Fibre Design and Fabrication

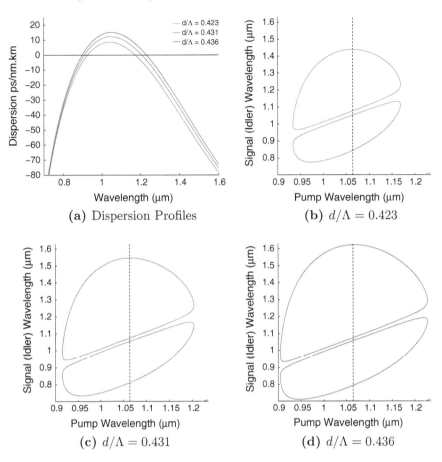

Fig. 4.3 Results of numerical simulations of the group-velocity dispersion (**a**), and resultant four-wave mixing phasematching contours (**b–d**), of a photonic crystal fibre with a fixed pitch of $\Lambda = 1.49\,\mu\text{m}$ and varying hole size: **b** $d/\Lambda = 0.423$, **c** $d/\Lambda = 0.431$ and **d** $d/\Lambda = 0.436$. Reducing the hole size pushes the dispersion profile further into the normal dispersion regime, reducing the separation between the zero-dispersion wavelengths. This allows us to control the phasematched wavelengths at the turning point of the idler branch without significantly affecting the orientation of the phasematching function

The finalised cladding structural parameters were $\Lambda = 1.49\,\mu\text{m}$ and $d/\Lambda = 0.431$ corresponding to a hole size $d \approx 0.65\,\mu\text{m}$. The dispersion profile, phasematching contours, phasematching function and JSI corresponding to these parameters are shown in Fig. 4.4. This combination of Λ and d/Λ correspond to a section of the phasematching contour where the contour lies at $+45^o$ on the signal branch and 0° on the idler branch. This yields an asymmetrically group-velocity matched state between the signal and pump fields, where the wavelength of the idler photon becomes less sensitive to changes in the pump wavelength. Provided that the bandwidth of the

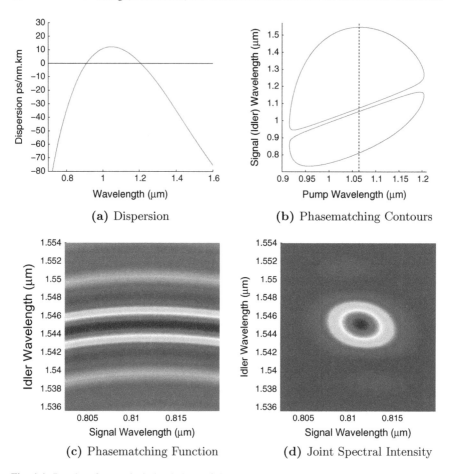

Fig. 4.4 Results of numerical simulations of the group-velocity dispersion (**a**) for the finalised PCF structural parameters, $\Lambda = 1.49\,\mu\text{m}$ and $d/\Lambda = 0.431$. From the dispersion profile the four-wave mixing phasematching contours (**b**), phasematching function (**c**) and joint spectral intensity (**d**) were calculated. At the central pump wavelength of 1064 nm, the phasematching function is orientated horizontally in $\{\omega_s, \omega i\}$, corresponding to a asymmetrically group-velocity state. A near circular JSI is achieved by matching the pump bandwidth to that of the phasematching function

pump envelope function is matched to the bandwidth of the phasematching function an uncorrelated two-photon state can be achieved.

4.2.2 PCF Fabrication and Characterisation

The selected PCF structure was fabricated with 8 rings of air holes using the stack-and-draw technique as shown schematically in Fig. 4.5. A large number of rings were

4.2 Photonic Crystal Fibre Design and Fabrication

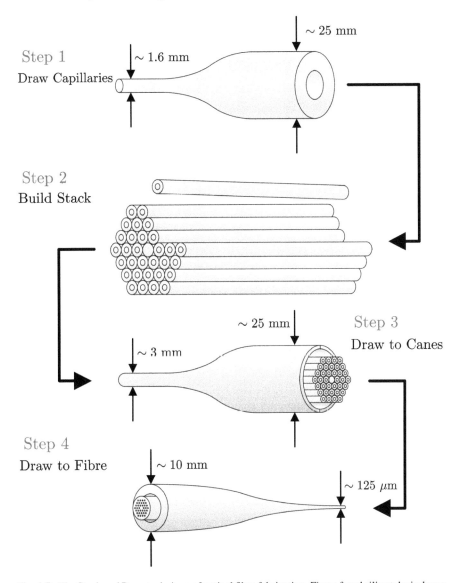

Fig. 4.5 The Stack and Draw technique of optical fibre fabrication. First a fused silica tube is drawn down to a diameter of a few mm. These capillaries are then stacked in a closed-pack hexagonal array in a specialist jig, in the *centre* of the stack a defect is incorporated; this *solid rod* will eventually form the core of the fibre. Once inserted inside the cladding tube, the stack is once again drawn down to a few mm to form a cane in which the air holes are preserved. Finally, the cane is inserted in the jacket tube and drawn down to 125 μm, whilst applying pressure to the air holes to keep them inflated

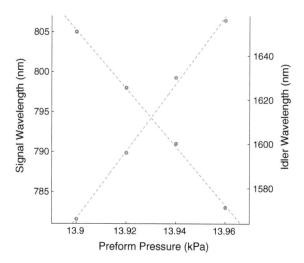

Fig. 4.6 An example calibration graph mapping the applied pressure during the draw to fibre and the four-wave mixing wavelengths, *dashed lines* correspond to a linear fit to the data points. We can use this graph to determine what pressure we need to apply to the cane to inflate the air holes to the correct size to achieve the target four-wave mixing wavelengths of 810 nm (*blue*) and 1550 nm (*green*)

used to ensure that the attenuation of the fundamental mode is low at long wavelengths. First, a fused silica glass tube of ∼25 mm diameter was drawn in the furnace at $2000^\circ C$, to approximately 1.6 mm in diameter to produce a set of capillaries each 1 m in length. These capillaries were arranged in a close-packed hexagonal pattern using a jig to produce a macroscopic scale preform of the desired PCF structure, this is called the stack. A solid fused silica rod was placed in the middle of the stack to form the core of the PCF. Next the stack was inserted inside a cladding tube and drawn in the furnace, whilst applying vacuum to the interstitial vacancies between the capillaries, to "canes", ∼3 mm in diameter. Each cane was inserted inside a jacket tube of outer diameter 10 mm, and drawn to ∼125 µm in diameter. Pressure was applied to the air holes of the cane to keep them inflated against surface tension so that they are maintained in the final fibre cross-section. Vacuum was again applied between the inner surface of the jacket tube and the outer surface of the cane to collapse the jacket onto the cane. Finally a polymer coating was applied to the fibre to stabilise it and to protect it from damage.

During the fibre draw, the structural parameters of the PCF can be controlled by varying the outer diameter of the fibre to set the pitch, and by changing the applied pressure to alter the size of the air holes. Various bands of fibre each with a different set of draw parameters and hence different cladding structures were drawn. At the beginning of each band, a sample was taken and the structure checked under an optical microscope for defects and to measure the pitch and hole size. However, due to the small size of air holes in this design, the microscope resolution was insufficient to accurately measure them during the draw. To ensure that the correct PCF structure was fabricated, the length of sample fibre was also pumped using a 1064 nm microchip laser and the FWM wavelengths observed on an optical spectrum analyser, see Fig. 4.7b. Figure 4.6 shows the variation in FWM wavelengths with applied preform pressure, this behaviour agrees with the simulated phasematching

4.2 Photonic Crystal Fibre Design and Fabrication

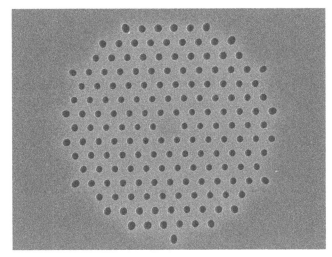

(a) Scanning Electron Micrograph

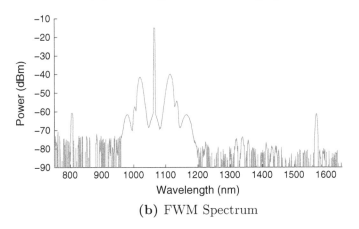

(b) FWM Spectrum

Fig. 4.7 *Top Panel* **a** Scanning electron micrograph of the fabricated PCF. In general, the air holes close to the core are regular in size in separation. In the outer ring the air holes are quite distorted, these imperfections should not impact the dispersion too heavily as the majority of the electric field is confined with the first few rings. *Bottom Panel* **b** Four-wave mixing wavelengths measured from a sample of fibre taken during the draw to fibre. The draw parameters can then be changed to generate the desired four-wave mixing wavelengths

contours in Fig. 4.3a. The draw speed and applied pressure were immediately adjusted to alter the structure for the next band. This process was then iterated until the desired wavelengths were generated from the PCF. The FWM wavelengths are very sensitive to changes in pressure. The smallest possible change that can be made in the applied pressure is 0.01 kPa. Therefore the magnitude of the pressure changes made during the draw was at the limit of the pressure handling system of the fibre tower. At the beginning of each band, 100 m of scrap fibre was drawn to allow the effect of the

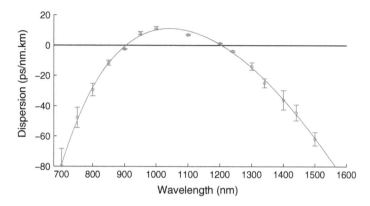

Fig. 4.8 Group-velocity dispersion profile of the fabricated PCF measured by *white* light interferometry with a Mach-Zender interferometer

pressure on the structure to stabilise. Once stabilised, the pressure was held constant throughout the band. The outer diameter of the fibre was continually measured during the draw, and found to vary around the set point by ±1.0%, due to fluctuations in furnace temperature and gas flow affecting the tension in the glass.

Following fabrication, the band of fibre whose phasematched FWM wavelengths most closely resemble the target was characterised further. This included obtaining scanning electron micrographs (SEM) of the PCF structure (Fig. 4.7a), measurement of the FWM wavelengths (Fig. 4.7b), group velocity dispersion by white light interferometry (Fig. 4.8) and the JSI through stimulated emission tomography (SET). Stimulated emission tomography is discussed in detail in Sect. 4.3.

4.3 Measuring the Joint Spectral Intensity Distribution

In Sect. 4.2, the target PCF cladding was designed to produce daughter photons that are group velocity-matched to the pump. When the pump bandwidth is set to match the length of fibre, a factorable joint spectral amplitude can be achieved. As we have seen in Sect. 2.4, the degree of spectral correlation present in the JSA can be determined by the Schmidt number K, obtained through the singular value decomposition of the JSA. To do this, we must have a method for experimentally determining either the JSA or JSI of the two-photon state generated in the PCF.

Measuring the full phase-dependent JSA for the two-photon state is very challenging from a technical standpoint. As a result of this, JSI is more commonly obtained through measurement instead. From the JSI, the JSA can be approximated by $f(\omega_s, \omega_i) = \sqrt{|f(\omega_s, \omega_i)|^2}$. However, this is in the absence of any phase information associated with the two-photon state. An inclusion of the phase will only ever reduce the purity of the state, therefore we can use the JSI to place a lower (upper) bound on the Schmidt number (purity) of the heralded state.

The most common experimental method for obtaining the JSI is by performing a spectrally resolved measurement of the coincidence count rate between the two-arms of the source [2–4]. The signal and idler outputs are sent to two independent spectrometers, the outputs of which are connected to single photon detectors. The coincidence count rate between these detectors is measured as the central wavelengths are scanned, mapping out the JSI for a pair of wavelengths at a time. This technique is used extensively, however the process is slow. This is due to the long integration times that are necessitated by the spectral binning of the measurement technique and the low pair-generation probability of the source. This constrains the potential resolution of the measurement, as a large number of counts are required for a low error, but the experimental running time must be short to reduce the potential for drift in the experiment conditions.

A second technique for obtaining the JSI is through the use of a fibre-based spectrometer. In this scheme, the signal and idler photons are coupled into long lengths of optical fibre, connected to the inputs of a pair of gated single photon detectors. In this case, the group velocity dispersion of the fibres maps different spectral components to different detection times. The gating times of the detectors can be scanned and the coincidence count rate mapped out in the detection times. By calibrating the system with bright light, the detection times can be converted into wavelengths and the JSI recovered. The lengths of the fibres and the temporal jitter of the detectors limit the resolution of the measurement. Both of the above methods have an inherent trade-off between resolution and running time that is due to the low probability of an event occurring from the spontaneous pair-generation process.

4.3.1 Stimulated Emission Tomography of FWM

A promising new technique was recently introduced by Liscidini and Sipe [5]. In this method, known as stimulated emission tomography (SET), it was shown that the JSI could be recovered through measurements of the stimulated forms of the non-linear conversion processes. This was first demonstrated for spontaneous PDC in a AlGaAs ridge waveguide, where the equivalent stimulated process of difference frequency generation (DFG) was measured [6]. Following from this, SET has been applied to a number of different source architectures, including stimulated FWM in standard birefringent optical fibres [7], silicon ring resonators [8] and silicon nanowires [9]. Although this method requires the use of a tunable light source at either the signal or idler wavelengths, the measurement can be made at high resolution in a short period of time using an optical spectrum analyser rather than single photon detectors. This is a clear advantage over the other techniques discussed above.

In both the stimulated and spontaneous forms of the pair-generation processes, be it PDC and DFG or seeded and unseeded FWM, the same phasematching and energy conservation relations dictate the wavelengths that can be generated. Therefore, the

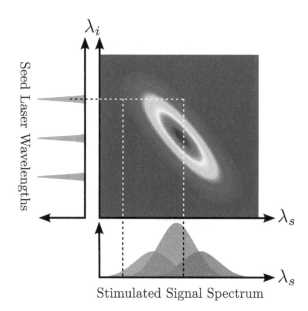

Fig. 4.9 An example of the JSI measurement. At each seed wavelength setting a "slice" through the JSI is taken, corresponding to the marginal signal distribution. By using many different seed wavelengths the entire JSI can be reconstructed from the set of marginal distributions

joint spectral distribution function of the spontaneous process becomes the response function for the stimulated form of the interaction. In the case of the spontaneous form of the interaction, where there is no classical seed field present at either of the daughter wavelengths, the process is reliant on vacuum energy ($\frac{1}{2}\hbar\omega_{s,i}$) to seed the conversion process and produce a pair. However, in the presence of a classical seed field at one of the daughter wavelengths, the amplitude of the signal and idler fields is greatly enhanced. This aspect is well known in classical non-linear optics, where it is widely used in optical parametric oscillators [10].

In the stimulated form of the pair-generation process, Liscidini and Sipe [5] showed that the average number of photons stimulated into the signal mode between ω_s and $\omega_s + \delta\omega_s$ is proportional to the joint spectral intensity distribution,

$$\langle \hat{n}(\omega_s) \rangle \delta\omega_s \propto |f(\omega_s, \omega_i)|^2 \delta\omega_s \delta\omega_i, \quad (4.6)$$

for a seed field with bandwidth $\delta\omega_i$ and central frequency ω_i. Therefore, by scanning a narrowband CW seed laser through the idler photon distribution, the spectral distribution of the stimulated signal photons can be mapped out. At each setting of the seed laser, a "slice" through the JSI, corresponding to the marginal signal spectrum is taken, which is recorded on the optical spectrum analyser, see Fig. 4.9. The set of recorded spectra for the stimulated photons in the signal mode can be used to reproduce the JSI. To this, the singular value decomposition is applied to determine the Schmidt modes and weightings. The purity is then calculated according to Eq. 2.32.

A schematic for the experimental set-up is shown in Fig. 4.10. The PCF sample is pumped at 1064 nm using the same *Fianium* fibre laser that was also used for pair-generation experiments in Chap. 6. The beam first passes through a half-wave

4.3 Measuring the Joint Spectral Intensity Distribution

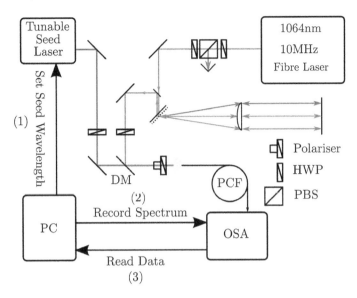

Fig. 4.10 A schematic of the experimental set-up used to perform stimulated emission tomography. The bandwidth and central wavelength of the pump pulses were controlled using a $4f$-grating spectrometer. The seed and pump beams are mixed on a dichroic mirror (DM) before being directed to the input of the PCF sample. Half-wave plates (HWP) and a polariser were used to co-polarise the two beams and select any polarisation axis in the PCF. The laboratory PC sets the seed wavelength before triggering a scan of the OSA, which when finished is read out by PC. This is then repeated for each new seed wavelength

plate (HWP) and polarising beamsplitter (PBS) for power control, before entering a $4f$-grating spectrometer that can be used to control the bandwidth and central wavelength of the pump pulses. For the seed field, a *INTUN-1550* tunable laser from *Thor Labs*, with a tuning range of 1500 to 1630 nm and bandwidth of <125 kHz was used. Both beams from the pump and seed laser pass through HWPs for polarisation control. The pump and seed beams are mixed on a dichroic mirror (DM) that is reflective for 1064 nm, with the seed beam passing through the rear. Both beams pass through a polariser; the polarisation state of both beams was then rotated, using the HWPs earlier in the set-up, to achieve maximum transmission through the polariser ensuring that they are co-polarised before entering the PCF. This arrangement of HWPs and polariser allow any polarisation axis of the PCF to be measured. Finally, the beams are coupled into the PCF using an aspheric lens and a 3-axis fibre coupling stage.

In Fig. 4.7a it can be seen that the PCF cladding structure does not possess perfect six-fold symmetry, as a result of this there will be a small amount of birefringence in the cladding. To provide a useful comparison between different samples of PCF, the polarisation state of the pump and seed beams must lie in the same direction through the cladding for all samples. During the fabrication stage, one of the air holes in the outer ring was replaced with a solid silica rod, to act as a marker. This was used to denote the orientation of the fibre at the launch optics. The input of each sample was placed in a bare fibre adapter (BFA) and examined using a fibre end viewer. The PCF was then rotated in the clamp, whilst observing the position of the marker, ensuring that marker was placed underneath the "key" of the BFA.

The output face of the PCF was mounted in a second BFA and connected directly to the FC-port of the *Ando AQ-6315B* spectrometer. A custom LabVIEW program was written for automatic data acquisition. Within the program, first the seed wavelength of the laser was set, once the laser has tuned to the correct position, the optical spectrum analyser (OSA) was triggered and a scan was run across the signal spectrum. Once completed, the spectrum was read out of the OSA over the GPIB connection to the PC. This was then repeated for the next selected wavelength. On each iteration, the new spectrum was appended to the next row of a matrix determining the JSI, where each row corresponds to one set point of the laser.

A typical measurement result is shown in Fig. 4.11. The spectral resolution along the signal axis is defined by the resolution of the OSA; here a resolution of 0.2 nm was used. Along the idler axis the spectral resolution is determined by the bandwidth and step-size of the seed laser, a typical step size of 0.2 nm was used; this could be reduced to improve the resolution. The central wavelength of the seed laser was calibrated by performing a scan over the range of idler wavelengths with no pump in place. All axes denoted idler, are calibrated to this measurement. A typical scan takes between 30 min and 1 hr, this is significantly faster than a spectrally resolved coincidence measurement, and could be improved by optimising the settings of the OSA. Through this technique, the effect on the JSI due to changes in the pump pulses can be seen quickly, allowing one to optimise the JSI for a specific section of fibre with relative ease. This fast turnaround in characterisation measurements is

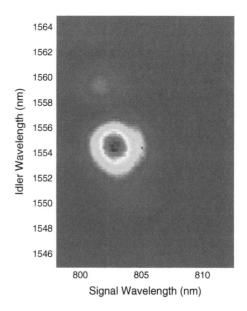

Fig. 4.11 An example of a typical JSI plot obtained from the stimulated emission tomography measurement. The OSA resolution was set to 0.2 nm, with a seed wavelength step size of also 0.2 nm. Overall, for this sample of PCF, a factorable state was achieved by matching the bandwidth of the pump to that of the phasematching function of the fibre. A low intensity secondary lobe at 1559 nm exists due to some small inhomogeneity in the fibre over this 30 cm length

4.3 Measuring the Joint Spectral Intensity Distribution

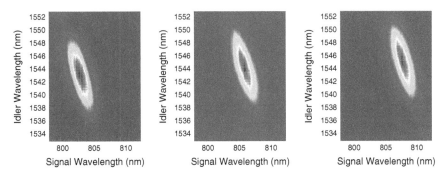

Fig. 4.12 Joint spectral intensity plots for the same PCF sample, recorded at 3 different central pump wavelengths. There is very little change in the idler wavelength as the pump wavelength is swept across the phasematching function. This indicates that the phasematching function is orientated to produce a asymmetric group-velocity matched state between signal and pump

particularly useful as it allows many different segments of the fabricated PCF to be sampled and measured.

Using stimulated emission tomography, a catalogue of characterised PCF segments can be produced. From this, pairs of fibres that exhibit identical JSIs were identified. Part of this process included determining the orientation of the phasematching function, and confirming that it is correctly positioned to allow asymmetrically GVM solutions between the signal and the pump. This was achieved by sweeping the pump envelope function across the phasematching function and determining the central wavelengths of the JSI. The central wavelength of the $4f$-spectrometer was changed and the stimulated JSI measured in turn. The resulting JSIs for three different central pump wavelengths are shown over the same range of wavelength in Fig. 4.12. Overall the marginal idler spectrum remains unchanged as the pump wavelength is swept across the phasematching function. The corresponding signal wavelength changes with the pump. This is a clear indication that the PMF is correctly orientated to achieve an asymmetrically group-velocity matched state between the signal and idler.

For long lengths of PCF, it was seen that many high intensity lobes appear in the JSI, distributed along the idler wavelength axis, see Fig. 4.13. The phasematching function for the asymmetrically group-velocity matched state, between the pump and signal, lies parallel to the signal axis in the JSI plots. Longitudinal inhomogeneity in the PCF structure has the capability to cause these features in the JSI. This would lead to fluctuations in the dispersion and hence variations in the phasematched wavelengths. This inhomogeneity poses a significant limitation on the length of usable fibre for photon pair generation and warrants further exploration. In the following sections we demonstrate through numerical simulations that these additional lobes are due to inhomogeneity and determine the length scale over which these dispersion fluctuations exist in the fabricated fibre.

Fig. 4.13 An example JSI showing significant inhomogeneity along the length of the PCF sample. In addition to the high intensity central lobe, a second lobe of moderate intensity arises due to a region of the sample with significantly different structural parameters. This degrades the purity that is achievable from the heralded state

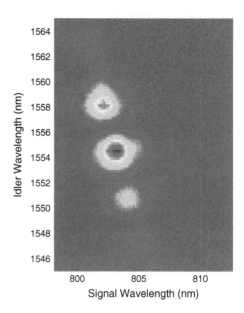

4.4 The Effect of PCF Inhomogeneity on the Reduced State Purity

During the process of deriving the form of the two-photon state in Sect. 2.2, it was assumed that the PCF was homogeneous along its length, and the phasematching function was calculated as,

$$\phi(\omega_s, \omega_i) = \chi^{(3)} \int_0^L dz \exp(i \Delta \beta z). \quad (4.7)$$

We can include inhomogeneity in the PCF by considering the fibre as being composed of m segments as illustrated in Fig. 4.14. Each segment is itself homogeneous with a length L_m, and is described by the parameters $\{d_m, \Lambda_m\}$ resulting in a phasemismatch of $\Delta \beta_m$. By performing the integration in a step-wise manner over each segment of homogeneous fibre, the overall phasematching function for the complete inhomogeneous fibre composed of m segments is [11]:

$$\phi(\omega_s, \omega_i) = L_1 \text{sinc}(\frac{\Delta \beta_1 L_1}{2}) \exp(i \frac{\Delta \beta_1 L_1}{2}) \\ + \sum_{n=2}^{m} L_n \text{sinc}(\frac{\Delta \beta_n L_n}{2}) \exp(i \frac{\Delta \beta_n L_n}{2}) \exp(i \sum_{l=1}^{n-1} \Delta \beta_l L_l). \quad (4.8)$$

The total phasematching function is a result of coherently combining the phasematching of each section by including the phase acquired on propagation through

4.4 The Effect of PCF Inhomogeneity on the Reduced State Purity

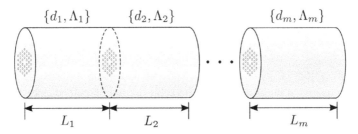

Fig. 4.14 An illustration of the inclusion of inhomogeneity in the PCF structure. Each sample of PCF can be broken down into a number of segments m of length L_m, each of which is described by a set of structural parameters $\{d_m, \Lambda_m\}$. The total phasematching function for the sample is found by integrating the phasemismatch step-wise along the total length of the fibre

the preceding sections [11]. This is achieved through the final multiplicative factor, $\exp(i \sum_{l=1}^{n-1} \Delta\beta_l L_l)$, on the right hand side of Eq. 4.8.

4.4.1 Numerical Reconstruction of Inhomogeneous PCFs

In order to determine the effect of inhomogeneity, the phasematching function in Eq. 4.8, was incorporated into the previous numerical model of the two-photon state and evaluated for a hypothetical fibre consisting of 3 different segments, with a total length of 1m. The structural parameters and length of each segment $\{d_m, \Lambda_m\}$ were chosen at random from a normal distribution centred on the mean values of $\bar{\Lambda} = 1.49\,\mu\text{m}$ and $\bar{d} = 0.64145\,\mu\text{m}$, with a variance of $\pm 0.5\%$ of the mean. The length of each segment was also selected at random, subject to the condition that the total length must be 1m. The structural parameters are shown in Table 4.1.

The JSIs of the independent homogeneous segments are shown in Fig. 4.15a–c, along with the JSI of the complete inhomogeneous fibre in Fig. 4.15d. From Fig. 4.15 it is clear to see that the inclusion of inhomogeneity degrades the factorability of the JSI due to the presence of multiple high intensity lobes. Even very small variations in $\{d, \Lambda\}$ within $\pm 0.5\%$ are enough to severely degrade the purity of the heralded state. In the actual fibre, these variations arise due to fluctuations in the draw parameters and

Table 4.1 Simulation parameters of an inhomogeneous PCF whose JSI is shown in Fig. 4.15d, including the reduced state purity of each segment

	Segment 1	Segment 2	Segment 3
d (μm)	0.64186	0.64345	0.64145
Λ (μm)	1.4904	1.4920	1.4900
L (m)	0.2670	0.2695	0.4635
P	82.7%	84.2%	89.6%

dimensions of the glass stock used during the fabrication process. In Fig. 4.6 it was seen that the phasematched wavelengths are particularly sensitive to variations in the applied pressure affecting the hole size. This is compounded by the lack of tolerance for small changes in the structural parameters within the current phasematching scheme.

From the results of these simulations, to achieve a pure heralded state, it is imperative that each length of fibre used in the final photon pair source is homogeneous. Next, the SET technique was used to identify the length scale over which the structural variations exist. This was achieved by cutting back through a long length of fibre, recording the stimulated JSI for each piece. The overall measurement "tree" is shown in Fig. 4.16. To begin with a 3 m long section of PCF was analysed, the stimulated JSI is shown in Fig. 4.17. There are clearly a number of regions within this length of PCF that exhibit different PMFs.

Next, the fibre was cut into three, 1 m sections and the JSI of each re-measured, see Fig. 4.18. There still appeared to be more than one phasematching function within each of the 1 m samples. Once again each of the 1 m sections was cut into three further sections each approximately 30 cm in length, producing in total nine 30 cm samples, and the JSI of each re-measured. A small selection of the measurements are shown in Fig. 4.19. On this length scale, some of the segments appear to be nearly homogeneous. Finally, we cut all nine 30 cm sections in half to produce eighteen 15 cm sections. At this length the fibres were too short to reach the spectrometer from the launch optics, and so an additional 30 cm of Hi1060 was spliced onto the PCF segments and once again the JSI was re-measured. Again only a small subset of the final 18 measurements are included here, see Fig. 4.20. In Fig. 4.20a, c vertical stripes appear running through the JSI. These are potentially due to interference caused by reflections off the splice joint and the end faces of the fibre or other optical elements in the set-up. At this length scale all of the sections are homogeneous, indicating that the dispersion fluctuations occur on a length scale between 15–30 cm. Therefore, when searching for usable pieces of fibre for the two photon pair sources, the fabricated PCF was sampled in lengths of 30 cm.

By using the model, including inhomogeneity, the phasematching in each of the measured segments can be numerically reconstructed. We can then work backwards from the 15 cm segments and reconstruct the full 3 m length. In doing so we can begin to quantify the degree of fluctuations present in the fibre over this length. To do this each segment of fibre is described by 3 parameters, d_m, Λ_m and L_m. We first assume that the majority of the dispersion fluctuations are due to variations in the pitch of the fibre, and fix d. This is justified, as during the draw the diameter monitor reports variations in the outer diameter of fibre due to fluctuations in the temperature, draw and feed speeds. These fluctuations are on the order of $\pm 1.0\%$ of the outer diameter. This in turn causes the fluctuations in the pitch of the fibre. Additionally, the holes in the PCF cladding are not identical in size. We select a value for d that when combined with values for Λ and L for the 15 cm segments produces the correct phasematched wavelengths.

For each segment we have a good starting estimate for L_m as at each stage the length of PCF was measured. For each 15 cm segment of PCF, the phasematching was

4.4 The Effect of PCF Inhomogeneity on the Reduced State Purity

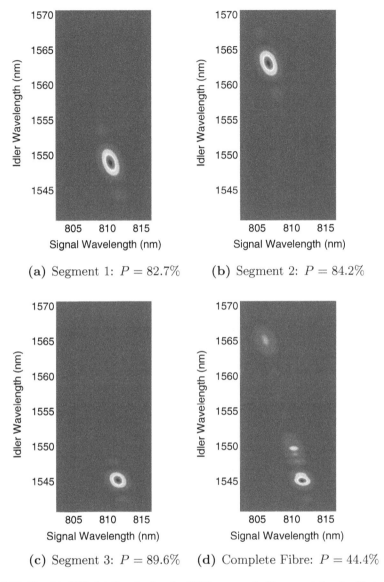

(a) Segment 1: $P = 82.7\%$

(b) Segment 2: $P = 84.2\%$

(c) Segment 3: $P = 89.6\%$

(d) Complete Fibre: $P = 44.4\%$

Fig. 4.15 Simulated JSI plots for a 1 m length of PCF composed of 3 segments (**a**–**c**), with structural parameters distributed within $\pm 0.5\%$ of the target mean, given in Table 4.1; **d** The JSI of the complete inhomogeneous PCF. By including even a small amount of inhomogeneity in the fibre the reduced state purity is severely degraded

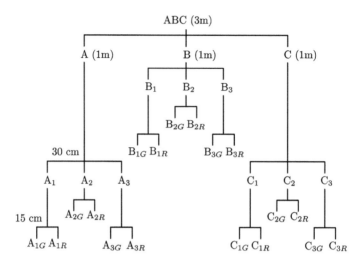

Fig. 4.16 A schematic of the measurement "tree" undertaken. The initial 3 m length (ABC) was cut into 3 pieces (A, B, and C) each 1 m in length. Each of these sections were cut into a further 3 segments, producing 9, 30 cm samples. Finally, each 30 cm segment was cut in half to produce a total of 18, 15 cm samples. The stimulated JSI of every segment was measured at each stage

Fig. 4.17 The stimulated JSI of the complete 3 m (ABC) PCF of sample. There are significant dispersion fluctuations over this length producing a JSI with many spectral features. Due to the long length of fibre, the bandwidth of the individual phasematching functions is very narrow

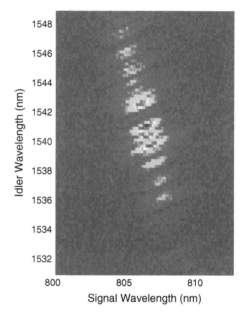

4.4 The Effect of PCF Inhomogeneity on the Reduced State Purity

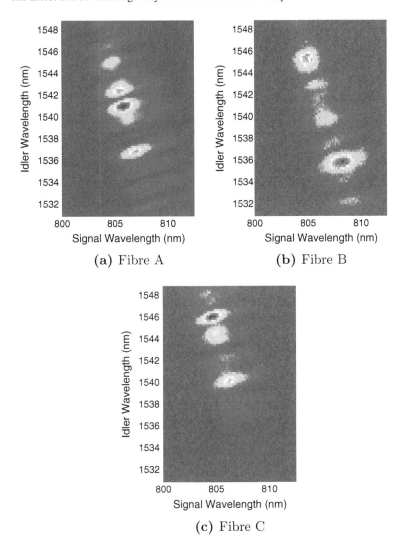

Fig. 4.18 The stimulated JSIs, for the three, 1 m length: **a** A, **b** B, and **c** C. The inhomogeneous dispersion along the length gives rise to several areas of different phasematching. As the fibre length is reduced the bandwidth of the phasematching function has broadened compared to the 3 m length

numerically reconstructed by keeping d fixed and using Λ as a fitting parameter to control the central phasematched wavelengths, and match them to the experimental measurement of the JSI.

With a starting set of parameters for each 15 cm segment, the 30 cm long segments were reconstructed in the inhomogeneous model. For each real 30 cm segment of PCF, we assume the inhomogeneity arises from three individual regions of phasematching, with a mean hole size. To perform the fit, the pitch was initialised using the values

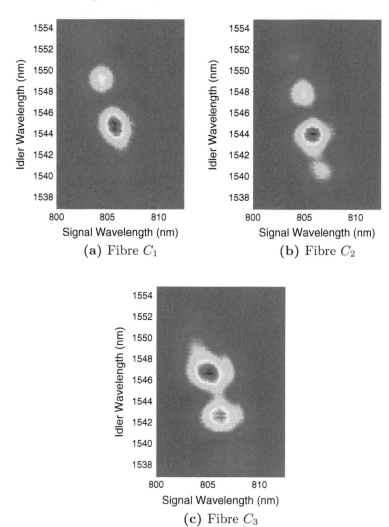

Fig. 4.19 The stimulated JSIs for the 3, 30 cm lengths cut from the sample C: **a** C_1, **b** C_2, and **c** C_3. At this length scale the inhomogeneity still persists for some of the fibre segments, also the bandwidth of the phasematching function has now become matched to the pump. In the absence of inhomogeneity, these samples would be capable of producing high purity heralded states

found from the reconstruction of the 15 cm segments that make up the 30 cm of PCF. The set of $\{\Lambda_1, \Lambda_2, \Lambda_3\}$ was then varied by small amounts to adjust the fit, and the changes in the positions of the high intensity lobes in the reconstructed JSI observed. We can define a minimisation function, $\mathcal{F}$,

$$\mathcal{F} = 1 - \sum_{\omega_s}\sum_{\omega_i} |f_{sim}(\omega_s, \omega_i)|^2 \cdot |f_{exp}(\omega_s, \omega_i)|^2, \qquad (4.9)$$

4.4 The Effect of PCF Inhomogeneity on the Reduced State Purity

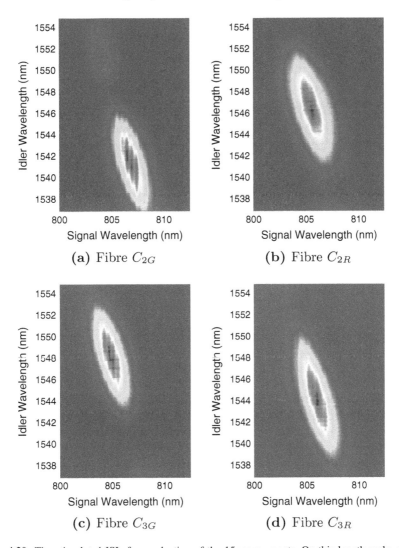

Fig. 4.20 The stimulated JSIs for a selection of the 15 cm segments. On this length scale, each segment is homogeneous, with a very broad phasematching function. In *panels* **a** and **c** the JSIs exhibit some vertical stripes, probably due to interference between stray reflections

where $f_{sim}(\omega_s, \omega i)$ and $f_{exp}(\omega_s, \omega_i)$ are the simulated and experimental JSIs respectively. By minimising $\mathcal{F}$, as $\{\Lambda_m\}$ and L_m are varied the fitting can be improved.

In the simple case, the 30 cm fibre segments often display two or three high intensity lobes in the measured JSI, one due to the first region of phasematching and another due to the second. The total length of fibre is fixed, within the numerical model we can then vary the lengths of the regions of individual phasematching slightly to achieve the correct relative contributions to the JSI. For example, if $L_1 > L_2$, then

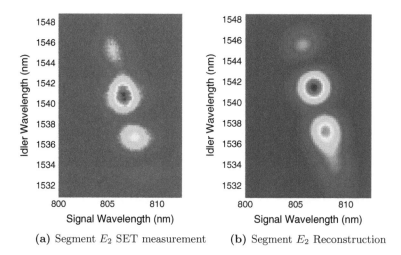

(a) Segment E_2 SET measurement (b) Segment E_2 Reconstruction

Fig. 4.21 An example of the numerically reconstructed JSI from an experimental measurement. **a** SET measurement of the JSI of one of the 30 cm segments. **b** Reconstructed JSI from the numerical model including inhomogeneity. The structural parameters of the model were adjusted to match the position and relative intensity of the features in the experimental JSI

the JSI is dominated by phasematching in segment one, the JSI will most closely resemble that of segment 1 with a small perturbation from segment 2.

This process is repeated for the remaining pairs of fibres until all nine 30 cm segments have been reconstructed. The reconstructed JSI of one of the 30 cm samples is illustrated in Fig. 4.21, together with the corresponding JSI measured from the experiment. Through this fitting method we have been able to achieve good agreement on the locations and relative magnitudes of the first three high intensity lobes in the reconstructed JSI. This process was repeated for the remaining eight 30 cm segments. Afterwards, each 1 m segment can be reconstructed from the three 30 cm segments it is composed from. As the fibre gets longer and more segments are included, the accuracy of each fit becomes more important to achieve good agreement between simulation and experiment. Therefore, at each stage we allow the fitting parameters to vary slightly once again, using the parameters fitted in the previous stage as a starting point. Again for brevity, we only illustrate the fitting for one of the three 1 m segments, see Fig. 4.22. Finally, the full 3 m of fibre was reconstructed from the three reconstructed 1 m segments, again we allow small changes in the fitting parameters to improve the overall fit, see Fig. 4.23. Overall the there is good agreement between the positions of the high intensity lobes in the JSI.

From the fitted values of Λ we can quantify how the pitch fluctuates along the length of the fibre. Figure 4.24 displays the fraction of the variance in the pitch $\Delta\Lambda$ to the mean $\bar{\Lambda}$, along the reconstructed fibre. In general, these fluctuations remain within the $\pm 1.0\%$ expected due to fabrication tolerances.

4.4 The Effect of PCF Inhomogeneity on the Reduced State Purity

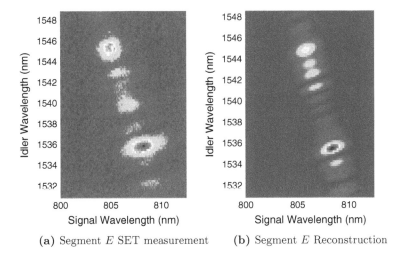

(a) Segment E SET measurement (b) Segment E Reconstruction

Fig. 4.22 An example of the numerically reconstructed JSI from an experimental measurement. **a** SET measurement of the JSI for a 1 m segment (B). **b** Reconstructed JSI from the numerical model including inhomogeneity. The structural parameters of the 30 cm segments (B_1, B_2, and B_3) that make up segment B, were used as the initial conditions for the fitting procedure. Small changes were then made in an attempt to improve the fit

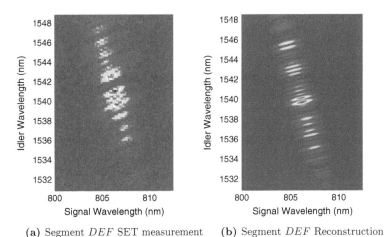

(a) Segment DEF SET measurement (b) Segment DEF Reconstruction

Fig. 4.23 An example of the numerically reconstructed JSI from an experimental measurement. **a** SET measurement of the JSI for the 3 m segment (ABC). **b** Reconstructed JSI from the numerical model including inhomogeneity. The structural parameters of the fitted 1 m segments (A, B, and C) that make up segment ABC, were used as the initial conditions for the fitting procedure. Small changes were then made in an attempt to improve the fit

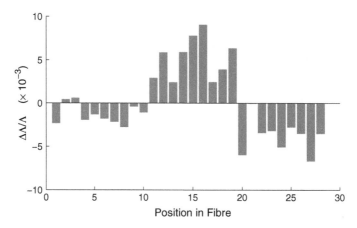

Fig. 4.24 Distribution of the variance in the fibre pitch, $\Delta\Lambda$, as a fraction of the mean $\bar{\Lambda}$, as a function of the position for the reconstructed 3 m fibre ABC. All values lie within $\pm 1.0\%$ of the mean, as expected from fabrication tolerances

4.4.2 The Final Fibres

In total two photon pair sources were constructed, for this two pieces of PCF are required. The two-photon state generated by each source must be factorable and identical. The SET technique was used to find two pieces of PCF that fulfill these criteria.

From the above study of the fabricated fibre, it was found that variations in the structure and fluctuations in the dispersion exist on a length scale of 15–30 cm. The remaining PCF was sampled in lengths of 30 cm and the stimulated JSI measured. From this a catalogue of more than 40 fibre samples was produced. In addition to possessing only a single piece of phasematching, the two pieces of fibre must be as identical as possible so that the photon pairs generated by each source are indistinguishable.

To identify pairs of fibres that most closely exhibit the same JSI, the overlap of the two JSIs was calculated. Each stimulated JSI was first normalised and followed by taking the square root, to approximate the JSA. The overlap integral,

$$\text{JSI Overlap} = \int\int d\omega_s d\omega_i \, f_{S1}(\omega_s, \omega_i) f_{S2}(\omega_s, \omega_i), \quad (4.10)$$

was then evaluated, where f_{S1} and f_{S2} are the approximated JSAs of source 1 and 2 respectively. Of the 40 samples characterised only 10 possess homogeneity over a length of 30 cm. From the remaining 10 samples, only two pairs of fibre were identified as being sufficiently similar from the degree of overlap. The stimulated JSIs of the selected fibre are shown in Fig. 4.25. Both fibres produce a factorable state with signal and idler wavelengths of 801 and 1560 nm respectively when pumped

4.4 The Effect of PCF Inhomogeneity on the Reduced State Purity

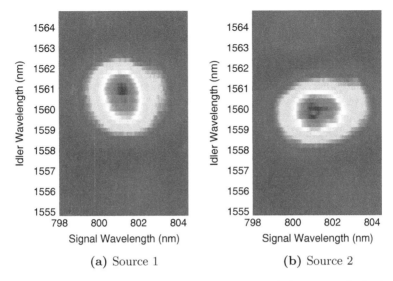

Fig. 4.25 Joint spectral intensity plots measured by stimulated emission tomography for the samples of PCF selected to be built into sources: **a** Source 1, and **b** Source 2. The *upper* bound on the reduced state purities, calculated using the Schmidt mode decomposition, were found to be $P = 86.2$ and $P = 86.9\%$ for source 1 and source 2 respectively. These two pieces of PCF show good agreement with a calculated overlap of 95% indicating that the photons from each source should be nearly indistinguishable

at 1058 nm. The purity of the reduced state for each fibre was calculated using the singular value decomposition and Schmidt mode decomposition on the approximated JSA with zero-phase, and was found to be 86.2 and 86.9% for source 1 and source 2 respectively. Finally, the overlap of the two fibres was calculated as 95% indicating that these two fibres are closely matched in terms of their output spectra.

4.5 Conclusions

In this chapter, the design, fabrication and characterisation of photonic crystal fibres for photon pair generation have been presented. A numerical model of the PCF was developed to calculate the phasematched FWM wavelengths and simulate the joint spectral amplitude and intensity functions. From this a fibre capable of producing a factorable joint state and hence pure heralded state was identified and fabricated using the stack-and-draw technique.

The PCF was characterised with a scanning electron micrograph of the cladding structure, white light interferometry to determine the group-velocity dispersion and an experimental measurement of the JSI obtained. To carry out the latter, the technique of stimulated emission tomography was introduced and a characterisation set-up constructed. Through this method, a catalogue of fibre samples was produced, from which two samples were selected to build into photon pair sources. An upper bound for the purity of the heralded state for each fibre was determined using the singular value decomposition of JSI, resulting in purities of $P = 86.2$ and $P = 86.9\%$ for source 1 and source 2 respectively. The overlap of the two JSIs was calculated as 95% indicating that the photons generated in the two sources possess a high degree of spectral indistinguishability.

The SET technique was also used to investigate the degree of inhomogeneity in the PCF cladding structure along its length. This manifests itself as additional features in the JSI, which reduce the purity of the heralded state. This effect was confirmed through a numerical simulation of an inhomogeneous PCF. By performing a series of cut-back measurements on a long length of PCF, and recording the JSI at each stage, the JSI was reconstructed in the numerical model to fit approximate structural parameters of the PCF in that section. From this, the degree of fluctuations was determined to lie within $\pm 1.0\%$ of the mean over a 3 m length, which is still large enough to degrade the two-photon state.

Finally, there is potential to develop an automated reconstruction process using a Monte-Carlo algorithm. An implementation of such a scheme is in development. If this could be achieved, then SET could shown promise for the characterisation of PCFs for use other than pair-generation.

References

1. K. Saitoh, M. Koshiba, Empirical relations for simple design of photonic crystal fibers. Opt. Express **13**, 267–274 (2005)
2. P.J. Mosley, J.S. Lundeen, B.J. Smith, P. Wasylczyk, A.B. U'Ren, C. Silberhorn, I.A. Walmsley, Heralded generation of ultrafast single photons in pure quantum states. Phys. Rev. Lett. **100**, 133601 (2008)
3. C. Söller, B. Brecht, P.J. Mosley, L.Y. Zang, A. Podlipensky, N.Y. Joly, P.S.J. Russell, C. Silberhorn, Bridging visible and telecom wavelengths with a single-mode broadband photon pair source. Phys. Rev. A **81**, 031801 (2010)
4. J.B. Spring, P.S. Salter, B.J. Metcalf, P.C. Humphreys, M. Moore, N. Thomas-Peter, M. Barbieri, X.-M. Jin, N.K. Langford, W.S. Kolthammer, M.J. Booth, I.A. Walmsley, On-chip low loss heralded source of pure single photons. Opt. Express **21**, 13522–13532 (2013)
5. M. Liscidini, J.E. Sipe, Stimulated emission tomography. Phys. Rev. Lett. **111**, 193602 (2013)
6. A. Eckstein, G. Boucher, A. Lemaître, P. Filloux, I. Favero, G. Leo, J.E. Sipe, M. Liscidini, S. Ducci, High-resolution spectral characterization of two photon states via classical measurements. Laser Photon. Rev. **8**, L76–L80 (2014)
7. B. Fang, O. Cohen, M. Liscidini, J.E. Sipe, V.O. Lorenz, Fast and highly resolved capture of the joint spectral density of photon pairs. Optica **1**, 281–284 (2014)
8. J.W. Silverstone, R. Santagati, M.J. Strain, M. Sorel, J.L. O'Brien, M.G. Thompson, Qubit entanglement on a silicon photonic chip. ArXiv e-prints (2014)
9. I. Jizan, L.G. Helt, C. Xiong, M.J. Collins, D.-Y. Choi, C. Joon Chae, M. Liscidini, M.J. Steel, B.J. Eggleton, A.S. Clark, Bi-photon spectral correlation measurements from a silicon nanowire in the quantum and classical regimes. Scient. Rep. **5**, 12557 (2015)
10. J.E. Sharping, M. Fiorentino, P. Kumar, R.S. Windeler, Optical parametric oscillator based on four-wave mixing in microstructure fiber. Opt. Lett. **27**, 1675–1677 (2002)
11. L. Cui, X. Li, N. Zhao, Spectral properties of photon pairs generated by spontaneous four-wave mixing in inhomogeneous photonic crystal fibers. Phys. Rev. A **85**, 023825 (2012)

Chapter 5
Construction of an Integrated Fibre Source of Heralded Single Photons

5.1 Overview

In this Chapter the construction of an integrated fibre photon pair source in PCF is described. The PCF fabricated in Chap. 4 was integrated with a wavelength division multiplexer (WDM) to split the signal and idler wavelengths into separate fibres. Fibre bragg gratings (FBG) are included at the pump wavelength to reject the pump light from the signal and idler fibres. To reduce the background count rates due to stray pump light and other photons occupying similar spectral modes, a length of photonic bandgap fibre (PBGF) was used, where the transmission bands of the fibre act as broadband filters. Generally, filtering is applied using interference filters in free space sources, and narrowband FBGs in fibre based sources. These filtering schemes are often required to produce a factorable two-photon state from a correlated one, discarding a large portions of the JSI and therefore potential heralded single photons in the process. We have designed the PCF cladding to produce a factorable state without the need for any filtering. Therefore, employing narrowband filter is only detrimental, and will reduce the achievable coincidence count rate. Throughout this thesis, these photonic bandgap fibres are designated PBGF-800 for the signal arm and PBGF-1550 for the idler. The design and fabrication of the PBGFs is discussed in detail in Sect. 5.2. In total, two sources of the same design were constructed. In Sects. 5.3 and 5.4 the assembly of the final sources and construction of the correlation electronics for characterisation is described. Finally, to multiplex the two sources a 2×1 optical switch is integrated with the idler arms, this is described in Sect. 5.5.

5.2 Photonic Bandgap Fibre Filter Design and Fabrication

Having generated some photon pairs the next stage is to ensure that only these photons reach the detectors and register a detection event. In reality however, along with the photon pairs, residual pump light and other spurious photons generated through a number of different processes are present in the guided mode of the fibre and must be removed before reaching the detector. Chief among these processes is spontaneous Raman scattering from the pump pulse. This generates noise photons most commonly in the long wavelength idler arm of the source. These uncorrelated singles counts degrade the quality of the single photon state and lead to the formation of a Poissonian number state instead. To prevent this, we must have a reliable manner with which we can spectrally filter the output of the source to remove any unwanted photons from the guided mode. However, this must be done with a broadband filter so that the factorable JSI is not perturbed. The level of isolation required between the pump and the signal or idler can be calculated by comparing the peak power of the pump laser pulse (av. power $\sim$1 mW at 10 MHz repetition rate) to that of a single photon at the signal or idler wavelengths. Assuming a pump pulse and single photon temporal duration of 2 ps, the level of isolation that is required from the filtering scheme is on the order of 100 dB.

In free space photon pair sources this can be achieved in a number of ways, such as using narrowband bandpass filters, coloured glass filters, monochromators, and spectrometers. In sources built within an integrated fibre architecture, filtering is much harder as everything must be accomplished in optical fibre. Previous all-fibre source architectures have made use of fibre-based components, such as optical circulators and FBGs, originally developed for the telecommunications industry [1]. By combing an optical circulator and FBG in series, see Fig. 5.1, only those wavelengths that lie within the transmission band of the FBG can travel from port 1 to 3 and remain in the guided mode. This scheme does however have some limitations. Firstly, the optical circulator typically has a high degree of loss ($\sim$1.5 dB per pass at 1550 nm, and often significantly worse at 810 nm). Secondly, the transmission window of the FBG is usually narrowband. As we have gone to significant effort to engineer a specific spectral state, we must avoid filtering out portions of it with a narrowband device. Finally, once the grating has been written, the central wavelength of the transmission band can only be marginally tuned through physically stretching the device. Before manufacture, one must know the wavelengths of the signal and idler photons with a good level of accuracy before committing to a device, especially as they can be quite costly to fabricate. In experiments this also removes the flexibility to tune the signal and idler wavelengths to achieve the best joint spectrum.

Here, we present a new filtering scheme which uses the transmission windows of an all-solid PBGF to act as a broadband filter. All-solid photonic bandgap fibres are another class of microstructured fibres, much like the PCFs discussed in Chap. 4. However, instead of inserting an array of air holes into the structure, the cladding

5.2 Photonic Bandgap Fibre Filter Design and Fabrication

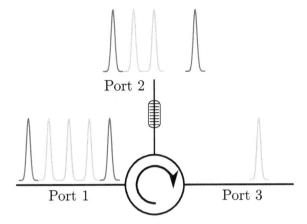

Fig. 5.1 A schematic of a filtering scheme using an optical circulator and a fibre bragg grating. Light entering port one can only travel clockwise round the device and so is directed through to port two. In port two a FBG with that has a high reflectance over a narrow wavelength range is inserted. This allows out-of-band light to pass through and exit port 2. The in-band light is reflected back into the circulator where it once again travels one further plot clockwise to the output. This allows one to filter out a desired spectral range from a broadband input

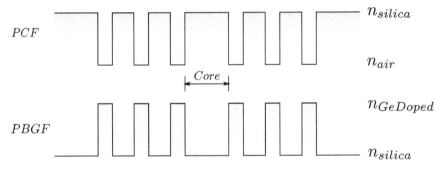

Fig. 5.2 *Top* refractive index profile for an air-clad PCF. As $n_{silica} > n_{air}$, light is confined by the traditional index-guiding mechanism. *Bottom* refractive index profile for a PBGF, where $n_{GeDoped} > n_{silica}$. Waveguides of this type cannot support index guided core modes, instead guidance is achieved through an optical bandgap formed due to the periodic high index lattice

is formed from inclusions of higher refractive index with a core of standard silica, see Fig. 5.2. This reversal of refractive indices between core and cladding means that the light in the core is no longer confined by total-internal reflection. The commonly accepted model for guidance in PBGFs (and other waveguides or similar structures) is the anti-resonant reflecting optical waveguide (ARROW) model developed by Litchinitser et al. [2].

Within the ARROW model, the origin of the high transmission bands can be understood by considering the sets of modes for the individual rods of the high index inclusions. In the case of a 2D PBGF, the rods that form the cladding can be thought of as single step-index cores. Each of these cores possesses a set of guided modes. If the wavelength of the light is below the cut-off wavelength of a rod mode, the light may be resonantly coupled out of the PBGF core and into the rod.

If we now consider the entire array of rods, their modes couple (hybridise) to form bands of supermodes. The transmission windows of the fibre core can then be explained in the behaviour of the cladding supermode bands as they approach cut-off. At certain wavelengths, above some cut-off threshold, a supermode of the cladding is no longer supported. Provided that there is not another supermode band present at this wavelength, then any light propagating in the PBGF core cannot be resonantly coupled out to the cladding, and so will remain confined in the fibre core. Over this wavelength range the cladding is said to posses a bandgap, a portion of the dispersion curve for which there is no allowed combination of frequency $\omega = ck$ and propagation constant β.

By changing the structural parameters of the cladding we can control the position and width of the bandgaps. Scaling d/Λ changes the width of the band whilst varying d alters the central position of the bandgap. By then tailoring these parameters we can design the transmission windows to fit to the FWM wavelengths. The task of design is then to determine the location of the band edges for sets of $\{d, d/\Lambda\}$. To do this a series of software packages developed Pearce et al. within the Centre for Photonics and Photonic Materials, at the University of Bath were used to calculate the photonic density of states (DOS) for the cladding, and from this find the locations of the bandgaps [3–5].

Straight away some simplifications can be made. The high-index glass stock available at the time of fabrication had a $d/\Lambda = 0.7$. This can only be reduced by jacketing the rod with extra silica glass, but this would widen the bandgaps potentially allowing more noise through to the detectors. Calculations were performed using structures normalised to the pitch of the fibre, thus the entire structure can be scaled simply by changing Λ. So even though the rod-rod separation does not actually affect the positions of the band-edges we can move them by scaling Λ.

For these calculations, an infinite periodic triangular lattice of high-index rods was used. Figure 5.3 shows the simulation unit cell and the calculated photonic density of states, for a cladding rods with $d/\Lambda = 0.7$ and refractive index contrast $\Delta n = 2.02\%$. Areas within the simulation cell marked in white correspond to the high-index inclusions where as those in black correspond to the lower index background. Within the DOS, supermodes of the cladding are shown in greyscale, where white regions correspond to areas of high DOS. The areas in red corresponding to regions where propagation in the cladding is not allowed are the bandgaps of the cladding. The blue horizontal line represents the light line of the simulation; this represents the target effective index of the core mode of the fibre. As the core is composed of

5.2 Photonic Bandgap Fibre Filter Design and Fabrication

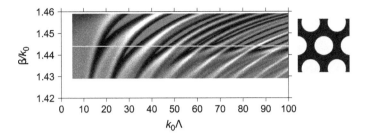

Fig. 5.3 *Left* Photonic density of states of the PBGF cladding. Areas of high DOS are shown in *grey scale*. The bandgaps, where there is no supermode of the cladding, are shown in *red*. The size of the rods, normalised to the pitch of the structure, controls the location of the bandgaps. Thus by scaling the pitch Λ, the bandgaps can be placed for a value of k_0 or λ. *Right* The simulation unit cell. Areas of high refractive index are shown in *white*; where as those regions in *black* correspond to the lower fused silica background

Table 5.1 Summary of the finalised structural parameters for PBGF-800 and PBGF-1550 and the final fibre outer diameters

Fibre	Λ	d/Λ	OD
PBGF-800	6.8 μm	0.7	~180 μm
PBGF-1550	13.5 μm	0.7	~265 μm

pure silica, this is set as the refractive index of fused silica at the target wavelength. The regions of high DOS above the light line correspond to the guided modes of the cladding. At the point of intersection between a band-edge and the light line the mode becomes cut-off and a bandgap opens up. For the target k_0 or λ that we wish to confine in the core mode, a value of pitch can be determined that places it within one of the bandgaps of the cladding.

From these calculations two designs were finalised, one for the signal photon (PBGF-800) and one for the idler photon (PBGF-1550). In each of these designs the third bandgap was targeted as it should experience less bend loss [5, 6]. The structural parameters of the two fibres are summarised in Table 5.1. The fibre was then fabricated using the same stack-and-draw technique as described in Sect. 4.2. A single stack was produced as the d/Λ remains constant in each fibre and from this two different sets of cane sizes were drawn. As with the FWM PCF fibre draws, the fibre was sampled off the drum and the transmission spectrum measured using a supercontinuum light source, the draw parameters were then adjusted to set the bandgaps in the correct positions. Figure 5.4 shows the PBGF transmission over a broad range of wavelengths.

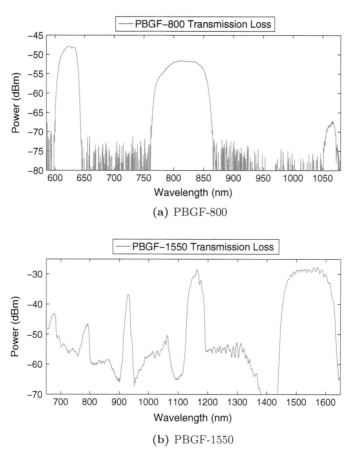

Fig. 5.4 Photonic bandgap fibre transmission spectra measured through approximately 2 m of fibre using a broadband supercontinuum light source coupled into the core. The output was collected with a multi-mode fibre and measured using an optical spectrum analyser. *Top* PBGF-800 shows a broad transmission band centred at 810 nm to collect the signal photon. *Bottom* PBGF-1550 shows a broad transmission band centred at 1550 nm to collect the idler photon

5.3 Component Assembly and Splicing

With the pair-generation PCF and PBGF successfully fabricated and the remaining components purchased such as WDMs (*Gooch and Housego*) and FBGs (*AOS*), the final assembly of the integrated photon pair sources was carried out. As all the components have been sourced in optical fibre this is achieved by fusion splicing.

The two fibres to be spliced are first cleaned and cleaved before being seated within a pair of clamps within the splicer. A set of motorised states can then be used to align the fibre cores (or cladding) and position the flat end faces close together. In a fusion arc splicer, an electrical arc between two electrodes of a controlled duration

5.3 Component Assembly and Splicing 101

and intensity is applied to heat and soften the glass. Simultaneously, the stages move together to fuse the glass at the interface between the two fibres. After splicing, a splice protector is placed around the joint and heat shrunk down around the fibres, producing a strong and stable join between the two fibres.

5.3.1 PCF-SMF Splicing

For conventional fibres (SMF28, Hi1060) of similar types, splice losses of < 0.1dB can be achieved. However, splices between micro-structured fibres and conventional single mode fibres (SMF) are often more difficult and, unless care is taken, can be very lossy. The reasons for this are two-fold. Firstly, if the mean field diameter (MFD) of the mode in each fibre are significantly different, then the modal overlap is poor and little light will be launched into the second fibre, resulting in a large loss at the interface. As we have designed the PCF with a particular $\{\Lambda, d/\Lambda\}$ to achieve a specific phasematching condition the mode field diameter is fixed.

Secondly, and particularly for air-clad PCFs, the waveguiding cladding structure near the splice can be damaged by the heat and stress applied in the splicing procedure. When heated, the air holes in the cladding will naturally want to collapse due to surface tension at the glass-air interface. This leads to a reduction in the refractive index contrast at the splice point. As a result of this, the MFD expands as it approaches the splice and may no longer overlap well with the guided mode of the next fibre [7, 8]. Alternatively, if the air holes are particularly small, such as in this PCF, the holes may collapse completely leading to loss of the waveguide in this region and high levels of loss. Therefore, to minimise the loss when splicing PCF it is important to tailor the arc current and duration to prevent hole collapse.

To achieve the lowest levels of splice loss the splicing arc can be used to gradually collapse the air holes in the cladding in a region up to the splice interface. If this is done in a sufficiently controlled manner then the two modes can be made to match [7–9]. This technique was not attempted here however as only short lengths of identical usable fibre had been identified. Any damage to the selected fibre segments may have resulted in them becoming unusable. Instead the splice settings were first optimised on similar fibres from another band until a loss of <50% was readily achieved. At this point the two selected PCF segments were spliced to 1 m lengths of Hi1060 and stored safely until use.

5.3.2 PBGF-SMF Splicing

The next type of splice to be considered is that between SMF and the PBGFs. For PBGF-800 this was relatively simple, the outer diameter and core size are similar to that of Hi1060 and SM800 and the large volume of silica glass at the interface means that a splice setting for SMF-SMF fibres could be used. For PBGF-1550 significant

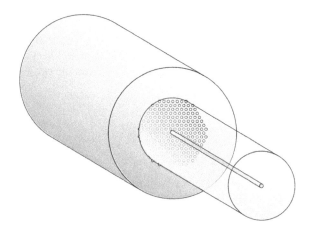

Fig. 5.5 Schematic illustrating the cladding overlap between PBGF-1550 and SMF-28. Due to the large physical extent of the PBGF cladding, there is very little overlap between areas of silica in the PBGF and SMF28. This makes the splice very brittle and results in a large failure rate of devices

problems arose. With laser light launched into the first fibre and the cores aligned in the splicer, a transmission of between 50–60% was achieved. But due to the large pitch and number of rings required the physical extent of the cladding area is very large. Thus, when splicing to standard SMFs, the cladding of the single mode fibre makes contact with the cladding rods in the PBGF, see Fig. 5.5. This coupled with the tension in the larger fibre as it is bent in the splicer, resulted in a large failure rate of splices when they were removed. To stabilise the joint at the splice a simple, single graded-core fibre with good mode-overlap with the PBGF, whilst also having significant overlap between areas of pure silica in the PBGF jacket, was fabricated. This fibre is designated as large mode area (LMA) fibre throughout this thesis.

To model the mode overlap between the two fibres, the mode of the PBGF-1550 was calculated using the same set of software packages used to determine the DOS, see Fig. 5.6a. A second model was then developed in COMSOL 3.5 to simulate the mode of a fibre with a graded-core of varying sizes, see Fig. 5.6b. The overlap integral,

$$\text{Overlap} = \int\int dxdy \Psi_{PBGF}(x,y)\Psi_{LMA}(x,y), \tag{5.1}$$

between the fundamental modes, $\Psi(x, y)$, was then evaluated to give an indication of potential splice performance. The results of the calculation, along with a data set for step-index fibre with the same index contrast as SMF28 is shown in Fig. 5.7. For the step-index fibre there is turning point around $18\,\mu\text{m}$, where beyond it the MFD is larger than in PBGF1550. For the graded-index core fibre, there is a clear trend towards larger core sizes as the mode field expands. This presents a new problem however. At these values of core size and index contrast, the LMA fibres are not single mode. This could potentially lead to photons being launched into higher order modes of the fibre, this would then cause large amounts of loss at the interface with a single mode fibre later. Additionally, the idler photons would no longer be heralded into a single spatial mode reducing the indistinguishability that is required from the heralded state.

5.3 Component Assembly and Splicing

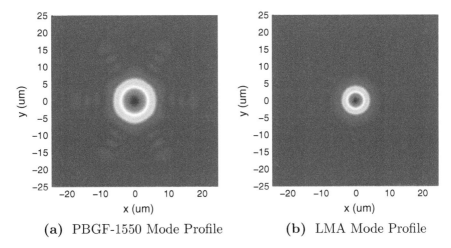

(a) PBGF-1550 Mode Profile

(b) LMA Mode Profile

Fig. 5.6 *Left* Fundamental mode profile of PBGF-1550 calculated using the FFPW eigensolver. The 6 lobes located around the perimeter of the central lobe exist in the high-index rods, and are a result of the resonance condition for the electric field. *Right* Fundamental mode profile of the LMA fibre fabricated to bridge between PBGF-1550 and SMF28. The mode profile was calculated using a graded-index profile in Comsol 3.5

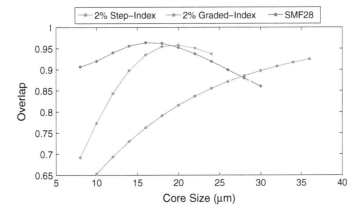

Fig. 5.7 Mode field overlap between the fundamental mode of PBGF-1550 and the fundamental mode of a large mode area fibre with a given core size. (*Blue*) 2% Step index fibre, (*Green*) 2% Graded Index Profile and (*Red*) A fibre with an index step equal to that of SMF28. For both, the 2% step and SMF28 profile fibres there is a clear peak in the overlap at a value around 18μm, above which the mode is now oversized. For the 2% graded-index fibre, the mode expands at a slower rate as the core size is increased and therefore the peak shifts to large core sizes

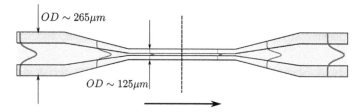

Fig. 5.8 A schematic of the taper profile of the large mode area fibre, the mode profile is shown in *purple*. As the fundamental mode propagates from the *left* to the *right*, the transverse structure is scaled down, provided that the change is gradual enough the guided mode will follow and also reduce in size to match the new core size, in which only the fundamental mode is guided. On the up taper if the transition is again gradual enough, higher order modes will not be reintroduced to the output

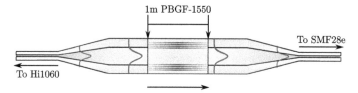

Fig. 5.9 A schematic of the complete PBGF filter device. The tapered LMA fibre is cut through at the waist to form two mode horns. The untapered ends are then spliced onto the PBGF, where there is now significant overlap of silica areas, providing a strong splice. The taper waists are spliced onto Hi1060 and SMF28, forming a device in which light can be transferred from SMF to PBGF and back again with relatively low loss

To prevent this from occurring the LMA fibres can be post-processed to incorporate a mode filter. This can be done by tapering the fibre down to produce a region with a smaller core size over which only the fundamental mode is supported at the target wavelength, see Fig. 5.8. Provided that the transition region is gradual enough the device will be low loss and the stripped modes will not be recaptured in the up-taper [10, 11]. After tapering the LMA fibres from an outer diameter of 265 μm to a diameter of 125 μm, the taper waist was cut in two, to form two LMA mode horns. Now the 265 μm ends can be successfully spliced onto the PBGF with lower loss, whilst the 125 μm waists can be spliced with low loss to conventional single mode fibres. A schematic of the final filter device is shown in Fig. 5.9. The fundamental mode of Hi1060 can be expanded to match the mode of PBGF-1550 in the up-taper, propagate through a length of PBGF-1550 to spectrally filter the generated photons, and then be reduced in the down-taper to that of SMF28. The transmission loss of the completed device is shown in Fig. 5.10b.

5.3 Component Assembly and Splicing

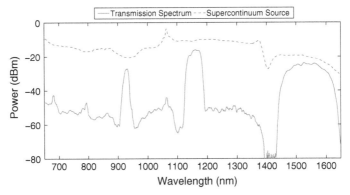

(a) Transmission Spectrum of PBGF-1550 Filter

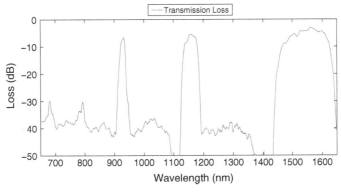

(b) Transmission Loss of PBGF-1550 Filter

Fig. 5.10 *Top* (*Blue*) Transmission profile of the complete PBGF-1550 filter measured by coupling supercontinuum light into the Hi1060 pigtail. (*Green*) Supercontinuum spectrum after cutting back in to the Hi1060 pigtail. *Bottom* The loss of the complete device, calculated by subtracting the transmission profile from the background supercontinuum spectrum. At 1550 nm the loss is approximately −3 dB

5.3.3 Assembly of the Photon Pair Source

A schematic of the source is shown in Fig. 5.11. The pair generation PCF was first spliced onto an FBG centred on the pump wavelength, with a bandwidth of 28 nm and a reflectivity of 99.9%, to reject the pump light. The transmitted light passes through into a WDM that is used to split the light into three different arms, the signal (810 nm), idler (1550 nm) and any residual pump (1064 nm) light. The transmission spectrum of the WDM is shown in Fig. 5.12.

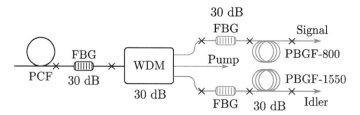

Fig. 5.11 A schematic of a single source. The 30 cm of PCF is splice onto the FBG written in Hi1060. The output of the FBG is spliced onto the input port of the WDM. In the 1060 nm output a connectorised Hi1060 patch cord was spliced, this can be easily coupled to a power meter. In the remaining two arms a further FBG was spliced on, followed by 1 m of PBGF-800 and PBGF-1550 for the signal and idler respectively. Each of the components provides 30 dB of isolation to the pump. In the signal arm, a connectorised SM800 patch cord was spliced onto the output to allow easy interface with the FC connectorsied detectors. The idler arm was *left* as bare fibre so that it can be integrated with the switch easily

The residual 1064 nm output fibre can then be coupled to a power meter to aid with coupling into the PCF and for monitoring the pump power level. In both of the signal and idler arms, a secondary FBG was first spliced onto the output fibres of the WDM to further increase the isolation of the pump. Following this, the two lengths of PBGF-800 and PBGF-1550 were spliced on to signal and idler arms respectively. After the filters, a length of single mode fibre, SM800 or SMF28 was spliced on to the signal and idler outputs respectively. Finally, the output of the signal arm in SM800 was connectorised with FC connectors to allow for consistent and easy coupling to the FC-port of the respective detectors. The final source was stabilised on a MDF board with laser cut channels to lay the fibres in. The total transmission loss from PCF to the three outputs of the source was then measured by performing a cut-back

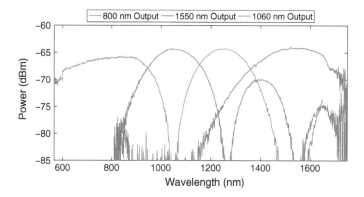

Fig. 5.12 Transmission spectra for the different outputs of the WDM. Supercontinuum light was coupled into the input port and each output port sent to the optical spectrum analyser. (*Blue*) 800 nm, (*Green*) 1060 nm, and (*Red*) 1550 nm. The higher order transmission bands can be seen for the 800 and 1060 nm output ports. The Si APDs are not sensitive to light in the higher band

5.3 Component Assembly and Splicing

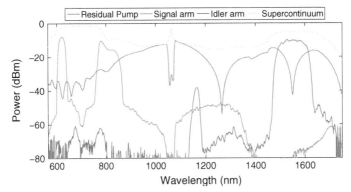

(a) Transmission spectra through complete device

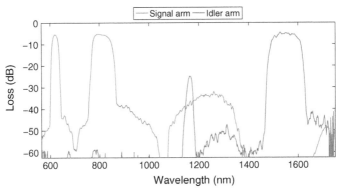

(b) Transmission loss through complete device

Fig. 5.13 *Top* Transmission spectrum of the complete source. Supercontinuum PCF was spliced onto the Hi1060 input rather than the pair generation PCF. This fibre was pumped at 1064 nm to generate a broad supercontinuum in the input fibre. Each of the outputs of the source was taken to a optical spectrum analyser, (*Blue*) 800 nm, (*Green*) 1060 nm, and (*Red*) 1550 nm. The Hi1060 input was then cut and taken to the optical spectrum analyser to measure the input spectrum (*Cyan*). *Bottom* The calculated loss from the cutback measurement for the (*Blue*) 800 nm and (*Red*) 1550 nm outputs, yielding a loss of −5.6 dB and −5.0 dB respectively

measurement over the complete device, see Fig. 5.13. The losses were found to be −5.6 dB, −27.3 dB and −5.0 dB for 810 nm, 1064 nm and 1550 nm respectively. The actual pump power coupled into the PCF can be found by using the loss at 1064 nm as a correction factor to the residual power in the output fibre. Overall an isolation to the pump in excess of 90 dB was measured using an OSA, this was limited by the dynamic range of the instrument. In both the signal and idler arms the PBGFs yield an isolation of ∼30 dB. The signal and idler arms also contains two FBGs at 1064 nm, each of which contributes 30 dB to the total isolation. Overall when combined with the WDM used to split the photon pairs from the pump, which itself contributes a further 30 dB, a total isolation in the region of 100–120 dB was achieved between the pump and the signal or idler wavelengths in the output fibres. The source is robust, low maintenance and can be easily taken off the optics table and stored or moved to a different laboratory, with no extra alignment effort required. In total two sources of the same construction were produced to be used in the final multiplexed device.

5.4 Detection and Coincidence Counting Electronics

With the photon pair sources constructed we can now generate some photons, but first we must establish a system with which to detect them. To perform this measurement we need to have a robust and accurate method for determining when both APDs fire simultaneously. By counting the resulting number of coincidences in a given time interval we can determine the heralded generation rate.

Further characterisation measurements, such as the degree of second order coherence and Hong-Ou-Mandel interference visibility necessitate the ability to determine up to four-fold coincidences including a simultaneous measurement of all possible lower level coincidences between the detectors. Developing the coincidence counting electronics required to do this forms a large step towards successfully demonstrating a heralded single photon source. In the past these correlations have been used measured using nuclear instrumentation modules (NIM) and precise lengths of coaxial cable to tune the delay between detector pulses. The cost and bulk of these systems limit the flexibility of the measurements that can be made.

An emerging trend in the single photon detection and correlation measurement is the use of field programmable gate arrays (FPGAs)[12]. FPGAs are programmable logic chips; fundamentally this allows one to program them to carry out a desired logical function, such as looking for coincidences. There is no limit to the number of times the FPGA can be reprogrammed; this makes prototyping of designs fast, with no soldering of components necessary to change design. If further correlation measurements are required the logical function can be quickly adapted, rather than having to purchase additional pieces of equipment. FPGAs are a powerful tool for signal evaluation conducted on a large scale at high speed. This makes these devices an ideal piece of hardware for photon counting applications, where the number of potential correlations scales exponentially with the number of detectors as well as operating with high repetition rate lasers. A further benefit is the low cost, a development board kit containing an FPGA, microprocessor, and variety of connectors (USB and Ethernet) can be purchased for $\sim$£100.

The FPGA system developed as part of this project was designed to have: 6 BNC inputs, capable of receiving digital pulses with durations of 35 ns (Si APD) and 100 ns (InGaAs APD), determine up to four-fold correlations between 6 inputs, count the number of all possible correlations observed and send the count rates off-board to a computer for data collection. A development board from Altera (DE2-115) was used to implement the design. A daughter card was fabricated with BNC connectors and an Ethernet jack, and attached to one of the expansion headers of the main development board.

Within the logical function developed for the FPGA, the input detector pulses are first synchronised to a fast (400 MHz) internal reference clock in a *clock-domain crossing*. In doing so, the pulses are shortened to 6 ns in duration and the rising edges of the signals locked to that of the clock. This time duration sets the coincidence time window of the correlation, i.e. if two signals both go through a rising-edge within 6 ns of each other they count as correlated. When the logical function is fitted to the logic cells within the chip the physical path taken by the signals between cells may cause timing delays between logically related signals. By locking all the inputs to a synchronous clock this timing skew can be minimised, ensuring that two correlated signals arrive at the correct portion of the design and register as a coincidence within the coincidence time window. Therefore the correlations can be determined very simply using sets of AND gates, where the output of the gate is logically high when both inputs are logically high. A series of counters monitor the output of the correlation functions and on every rising edge of the output increment by +1.

The value of the counters are read out once per second and sent from the FPGA to the laboratory PC via a local Ethernet connection. A LabVIEW application on the PC-side monitors the incoming connection, reads the data out of the Ethernet frame and displays it on screen. The count rates for every correlation are continuously written to a text file that is then saved at the end of a data run.

5.5 Optical Switch Integration

To complete the multiplexed device, the idler outputs of the two sources were connected to the inputs of a 2×1 fibre coupled optical switch. Here, a Nanona™ Opto-Ceramic™ switch from Boston Applied Technologies Inc. was used. The behaviour of the switching scheme is described as follows. With the switch powered on, if the switch is closed, the output from source 1 is routed through to the common output, see Fig. 5.14a. This is the default position. The heralding detector of source 2 was connected to the switch set pin. Thus, when the heralding detector of source 2 fires, the switch opens and routes source 2 to the common output, see Fig. 5.14b. The switch itself operates by means of the electro-optic (EO) effect. A voltage, ($V_\pi = 199$ V), is applied longitudinally to the ceramic in the direction of light propagation. A series of micro-patterned electrodes on the front face produce a spatially varying electric field transverse to the direction of light propagation. Through the EO-effect, this introduces a refractive index gradient in the ceramic which is used to steer the beam, and map switch inputs to outputs [13]. Unlike a mechanical switch, this allows the switching rise time (set-up time) to be fast at around 60 ns (mainly limited by the switch driver electronics). Factoring in the additional set-up time due to the electronic switch drive unit, the switch can be operated at a high repetition rate, specified

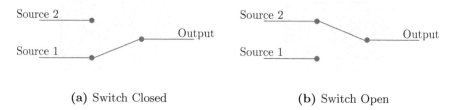

Fig. 5.14 Illustration of the switch states: **a** Closed—with no heralding signal from Source 2, the switch remains closed and routes Source 1 to the output. **b** Open—when a heralding signal is present from Source 2 the switch opens and routes Source 2 to the output

as 1 MHz by the manufacturer. Combined with the low insertion loss (<1 dB) and polarisation insensitivity make this a good choice of switch.

The time that the switch is open to source 2 is approximately equal to the duration of the 5 V pulse placed on the set pin. The detector pulses from the Si APD heralding detector are 35 ns in duration, and therefore too short to allow the switch to fully open. To ensure that the maximum switch transmission is achieved, the heralding pulses were first sent to a second FPGA. Within this FPGA, the logical function stretches the input pulses from 35 to 140 ns, incurring a small amount of additional time delay (45 ns) in the process, see Fig. 5.15a. The output pulses from the FPGA are used to drive the set line of the switch. The additional delay in the FPGA was adjusted to match the arrival time of the idler photons from source 2, placing the photons in the middle of the switching window, where hopefully the transmission is at a maximum. Figure 5.15b displays the transmission window of the switch, with the beam from a CW 1550 nm laser coupled into one of the input fibres. The switch driver causes an additional set-up delay of 150 ns. The total set-up time between a heralding signal being generated and the switch opening is around 200 ns.

In the characterisation measurements, we wish to detect the coincidences between the signal and idler photons. To do this the electronic pulses from the detectors must arrive at the FPGA-Correlator simultaneously. However, now the idler photons are detected around ∼200 ns after the signal photons, so we must delay the heralding signals in time by this amount. This is an impractically large a time delay to be done with only lengths of coaxial cable. Instead, the same FPGA that produces the switching signal, also produces a variable electronic delay for the heralding detector signals before they are routed to the correlator, Fig. 5.15c.

A schematic of the complete multiplexed system is shown in Fig. 5.16. When integrating the switch with the multiplexed system, the idler photons from both sources must be delayed by the total set-up time of the switch so that the state of the switch is prepared ahead of the arrival of the photons. To accommodate the approximately 200 ns of set-up time, 42 m of SMF28 fibre was spliced onto the output of each source before splicing to the switch itself. A pair of polarisation controllers were added to the fibre delay line to control the polarisation state and counter any polarisation effects induced by propagating through the fibre wound on a pair of spools.

5.5 Optical Switch Integration

Fig. 5.15 Testing the FPGA controlled switching scheme, in all panels the input trigger (heralding) signal is shown in (*blue*). **a** As the switch and drive unit have a rise-time (set-up) of 60 ns, the FPGA first produces the 140 ns switching signal (*red*) from the 35 ns heralding signal (*blue*). This allows the switch to reach point of maximum transmission. **b** The optical profile of the switch window (*red*). Measured with a fast photo-diode and CW laser, and the switching pulse generated in (**a**). The switch driver causes a further 150 ns of set-up delay. **c** The same FPGA is also used to delay both heralding channels by 200 ns (*green*), so that the heralding and idler detector signals reach the FPGA-correlator simultaneously

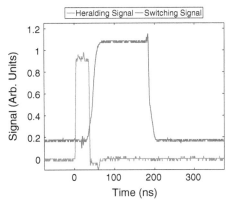

(a) Switch set signal

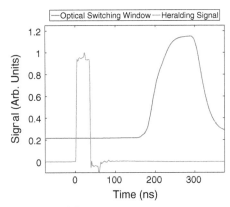

(b) Switch optical output

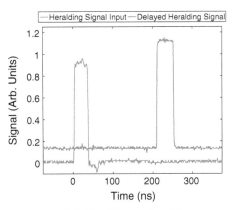

(c) Heralding pulse delay

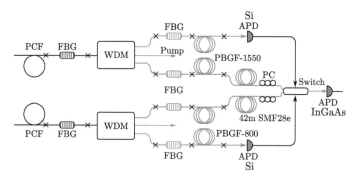

Fig. 5.16 Schematic of the all-fibre multiplexed source. The PCF is first spliced onto a FBG (*Black*), to reject the pump light. The signal and idler photons are split in the WDM, before passing through secondary FBGs to increase the isolation to the pump. The photons next pass through lengths of PBGF to filter out any remaining uncorrelated photons. Polarisation controllers (PC) on the idler fibres are used to match the polarisation of the photons at the output. The heralding detector of source 2 sets the state of the switch

In addition to ensuring the switch is prepared ahead of the arriving photons, the idler photons from the two sources must be made to occupy the same time bin at the InGaAs detector. As a pulsed laser is used the idler photons are localised in time to within the pump pulse duration. Therefore, to reduce the background count rates at the detector, it is run in a gated configuration where the element is only active for a short period of time, centred around the expected arrival time of the idler photons. The required gating signal is derived from a photo-diode placed in a beam dump of the pump laser pulse train. Within the InGaAs detector the internal delay between the arrival of the gating signal and the activation of the detector bias voltage can be controlled to encompass the arriving photons. The relative offset between the arrival times of idler photons from each source was measured by pumping both sources, initially independently, with the switch set statically for each source in turn. The internal trigger delay of the InGaAs detector was stepped through and the detector count rate integrated for 30 seconds at each point, the resulting arrival times are shown in Fig. 5.17. Initially, source 2 was found to be marginally ahead by 5.1 ns, a further 1.24 m of SMF28 was spliced in between source 2 and the switch to introduce the required additional delay. Further to this, a variable delay line using a translation stage and a pair of mirrors was constructed for the pump beam to source 1. This made it possible to fine-tune the timing of the pump pulses, and hence the arrival time of the idler photons at the detector to within the gate width.

As a final check, both sources were pumped simultaneously with the switching scheme active and the count rates observed. Each source was blocked in turn and the count rates measured. With source 1 and the switch active, the count rates were found to not change significantly from the static case, this is as expected as this is the default position of the switch and is in effect static. With source 2 and the switch active, the number of raw idler counts at the InGaAs detector was significantly reduced,

5.5 Optical Switch Integration

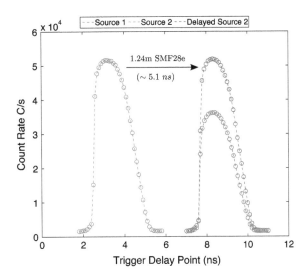

Fig. 5.17 The time of arrival of idler photons from source 1 (*green*) and source 2 (*blue*), relative to the detector trigger signal. Measured by scanning the internal trigger delay and counting the number of detection events. Source 2 was initially found to be 5.1 ns ahead of source 1, a further 1.24 m of SMF28 was spliced in to source 2 to add required delay. The count rate of source 2 was re-measured to confirm that the arrival times of idler photons from the two sources were in the same time bin (*red*)

but crucially the number of coincidences with the heralding signal remained nearly constant. This is due to a noise gating effect, which will be discussed in more detail in Sect. 6.2.3.

Now that both photon pair sources have been constructed and characterised classically, and we have a system with which we can count coincident detection events we can begin to characterise the quantum states generated. In the following Chapter we will discuss the results of the characterisation measurement carried out in the quantum domain.

References

1. A.R. McMillan, J. Fulconis, M. Halder, C. Xiong, J.G. Rarity, W.J. Wadsworth, Narrowband high-fidelity all-fibre source of heralded single photons at 1570 nm. Opt. Express **17**, 6156–6165 (2009)
2. N.M. Litchinitser, S.C. Dunn, B. Usner, B.J. Eggleton, T.P. White, R.C. McPhedran, C.M. de Sterke, Resonances in microstructured optical waveguides. Opt. Express **11**, 1243–1251 (2003)
3. J. Pottage, D. Bird, T. Hedley, J. Knight, T. Birks, P. Russell, P. Roberts, Robust photonic band gaps for hollow core guidance in PCF made from high index glass. Opt. Express **11**, 2854–2861 (2003)
4. G.J. Pearce, T.D. Hedley, D.M. Bird, Adaptive curvilinear coordinates in a plane-wave solution of Maxwell's equations in photonic crystals. Phys. Rev. B **71**, 195108 (2005)
5. T.A. Birks, G.J. Pearce, D.M. Bird, Approximate band structure calculation for photonic bandgap fibres. Opt. Express **14**, 9483–9490 (2006)
6. J.M. Stone, G.J. Pearce, F. Luan, T.A. Birks, J.C. Knight, A.K. George, D.M. Bird, An improved photonic bandgap fiber based on an array of rings. Opt. Express **14**, 6291–6296 (2006)
7. L. Xiao, M.S. Demokan, W. Jin, Y. Wang, C.-L. Zhao, Fusion splicing photonic crystal fibers and conventional single-mode fibers: microhole collapse effect. J. Lightwave Technol. **25**, 3563–3574 (2007)

8. L. Xiao, W. Jin, M.S. Demokan, Fusion splicing small-core photonic crystal fibers and single-mode fibers by repeated arc discharges. Opt. Lett. **32**, 115–117 (2007)
9. X. Zhou, Z. Chen, H. Chen, J. Hou, Fusion splicing small-core photonic crystal fibers and single-mode fibers by controlled air hole collapse. Opt. Commun. **285**, 5283–5286 (2012)
10. J. Love, W. Henry, Quantifying loss minimisation in single-mode fibre tapers. Electron. Lett. **22**, 912–914 (1986)
11. T. Birks, Y. Li, The shape of fiber tapers. Lightwave Technol. J. **10**, 432–438 (1992)
12. R.C. Pooser, D.D. Earl, P.G. Evans, B. Williams, J. Schaake, T.S. Humble, FPGA-based gating and logic for multichannel single photon counting. J. Mod. Opt. **59**, 1500–1511 (2012)
13. Q.W. Song, PLZT ceramic based electro-optic beam steering device for optical interconnection and pathway adjustment, Technical Report, Syracuse University (1996)

Chapter 6
Characterisation of a Multiplexed Photon Pair Source

6.1 Overview

In this Chapter we will build on the work presented in Chap. 5 and characterise the individual sources and the complete multiplexed device. First, in Sect. 6.2, the individual sources and multiplexed device are characterised in terms of the coincidence-to-accidentals ratio between the heralding and idler arms. Secondly, the use of the switch as noise gate for Source 2 is explored in Sect. 6.2.3. Thirdly, in Sect. 6.3.1, the marginal second-order coherence is measured to determine the spectral purity of the idler photons. Finally, a measurement of the second order coherence of the heralded idler state was used to demonstrate that the sources are heralding single photon states, see Sect. 6.3 for details.

6.2 Characterisation of the Coincidence Count Rates

Following from Sect. 5.5, where the idler photons were aligned into the same time bin of the detector, Source 1, Source 2 and the multiplexed system were fully characterised in terms of the coincidence count rates, coincidence-to-accidentals, marginal second-order coherence and heralded second order coherence. To select Source 1 individually, the switch was powered on but the set signal disconnected, to select Source 2 the switch was not powered on at all, and when multiplexing the switch is triggered by the heralding detector of Source 2. In addition to using the switch to multiplex the two sources, it could also be used as a noise gate when running Source 2 independently. In this case the switch is powered on with the set signal derived from the heralding detector. The pump beam to source 1 is blocked, therefore only when the heralding detector of source 2 fires will the switch open to the idler photons. As a result of this the amount of noise from those pump pulses that did not result in a pair-generation event is greatly reduced [1].

As the heralding detector is an integral part of the source, to maximise source performance, it is important to select a detector with high detection efficiency. The signal photons, around 800 nm, were detected using a four-way silicon APD from *Excelitas* (SPCM-AQR-4C), with a detection efficiency of ∼50% at the target wavelength.

The long-wavelength idler photons, around 1550 nm were detected using *ID210* and *ID201* InGaAs detectors from *idQuantique*. The user configurable detection efficiency was set to 15% to minimise the effect of detector dark counts. There are less stringent requirements on the idler detector as it is not part of the source and is only used for characterisation purposes in these experiments. Any inefficiency in the detector can be accounted for in observed count rates. It does however make the characterisation measurements harder to carry out, requiring a longer integration time to achieve smaller error bars as well as lacking sensitivity.

In all of the following characterisation experiments the Si APD was run in free-running mode, and so is able to detect a photon regardless of the arrival time (provided it is not within the dead time of a previous detection). The detector itself operates as a binary device. Therefore, any detection result constitutes a heralding signal, and thus the detector is prone to providing spurious heralding signals that are not correlated with events in the idler arm.

The idler detectors were run in a gated configuration from a signal generated from the pump laser pulse train. We can then set the gate time of the detector, such that the detector element is only active for a fraction of the laser pulse train that encompasses the expected time of arrival of the idler photons. By setting the gate width to 2.5 ns, a large portion of the noise reaching the detector can be removed from pulse train. After a detection event the signals are sent to the FPGA correlator via lengths of coaxial cable and a variable delay unit from *Ortec*. This variable delay can then be scanned to account for any differences in optical path length or other electrical delay, that causes a time delay between correlated detector signals arriving at the FPGA.

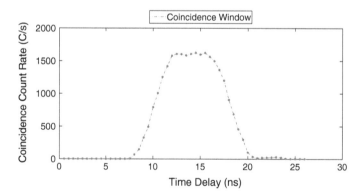

Fig. 6.1 Coincidence counts between a heralding detector and the idler detector as the electrical delay between the two signals is varied. The peak in coincidence counts between the two detectors indicates that the detector signals are due to the presence of a correlated generation process. The width of the peak is related to the coincidence time window in the FPGA-correlator

6.2 Characterisation of the Coincidence Count Rates

As the delay is scanned, the point of optimal overlap is found at the peak coincidence count rate. An example scan is shown in Fig. 6.1. For the different characterisation experiments, different amounts of delay will be required to align the detection signals in time at the FPGA-correlator.

6.2.1 Coincidence Counting and Coincidence-to-Accidentals

First the coincidence count rates between the heralding detectors and the idler detector for each individual source and the complete multiplexed device was measured as a function of pump power. From this, the coincidence-to-accidentals ratio (CAR) which is an estimate of the signal-to-noise of true coincidence counts from a correlated pair-generation event, and those coincidences observed by accident. An accidental coincidence is one in which the two detector signals result from uncorrelated processes. These events may occur from a number of different mechanisms, such as dark counts and after-pulsing in either detector, stray room light, residual photons at the pump wavelength, and photons generated from undesired non-linear processes such as spontaneous Raman scattering. Any combination of these events with a true signal or idler photon detection event may occur or as an isolated combination of noise processes.

Detector dark counts and after-pulsing effects were minimised by working at a low detector bias voltage (efficiency) and by introducing a dead time in the detector electronics. The effect of stray room light was minimised by shielding the detectors in a box lined with blackout fabric and working in a dark laboratory. By including a number of FBGs at the pump wavelength and lengths of PBGF filters there was >90 dB of isolation to the residual pump light and Raman scattered photons in the signal and idler arms.

A measure for the accidentals count rates can be determined by delaying the idler photons or the detector signal, to coincide with a signal detection event from the next pulse of the laser. We could then measure the cross-correlation count rate between a detection event in the idler arm on the n'th pulse and a signal detection event on the $(n+1)$'th pulse.

Another method of determining the accidentals count rate is to approximate it by calculating the probability that a number of detection events randomly distributed in time happen coincidentally [2]. With no time-stamping information available, we have no knowledge of which laser pulse resulted in a detection event within the integration time. From the number of detection events at each detector during the 1-second integration period, the probability per pulse of a detection event occurring is,

$$p_{sig} = \frac{N_{sig}}{R_p}, \qquad (6.1)$$

$$p_{idl} = \frac{N_{idl}}{R_p}, \qquad (6.2)$$

where N_s and N_i are the singles count rates for the signal and idler respectively and R_p is the pump repetition rate. The probability per pulse that the signal and idler detectors fire coincidentally, (p_{acc}), is found from the overlap of the two events,

$$p_{acc} = p_{sig} \cdot p_{idl} = \frac{N_{sig}}{R_p} \cdot \frac{N_{idl}}{R_p}. \tag{6.3}$$

We can convert this probability back into a count rate by multiplying by the repetition rate. Hence, the accidentals count rate per second is,

$$N_{acc} = p_{acc} \cdot R_p, \tag{6.4}$$

$$= \frac{N_{sig} \cdot N_{idl}}{R_p}. \tag{6.5}$$

The coincidence-to-accidentals ratio is then,

$$\text{CAR} = \frac{N_{coinc} - N_{acc}}{N_{acc}}, \tag{6.6}$$

a CAR in excess of 10 is an indication of a potentially useful source of heralded single photons [3].

6.2.2 Experimental Results: Coincidences and CAR

Figure 6.2, shows the general experimental set-up. The 10 MHz, 1064 nm, 200 fs pulses from the *Fianium FemtoPower* fibre laser first pass through a HWP and PBS pair for power control. A *DET10* photodiode from *Thor Labs* was placed in the rejected arm of the PBS, the signal from which was used as the gate signal for the idler detectors. Following the PBS, the laser train passes through a 4-f spectrometer. A transmission grating was used to spatially separate the different spectral components of the pulse. In the focal plane of a lens placed one focal length away from the grating, a mirror was placed to direct the beam back over the outgoing beam, through the filter to the grating. A pair of razor blades mounted on translation stages just in front of the mirror, can be used to select a portion of the spectral content of the pulse, setting the central wavelength and bandwidth of the pump pulses. The selected light passes backwards through the filter to the grating where the pulse is recombined. An optical spectrum analyser and autocorrelator were used to calibrate the filter in the spectral and temporal domains, the calibrate root mean square (RMS) bandwidth σ_p^{RMS} of the pump spectrum is shown in Fig. 6.3. A HWP before the 4f-spectrometer was used to adjust the polarisation of the pump to maximise the efficiency of the grating.

6.2 Characterisation of the Coincidence Count Rates

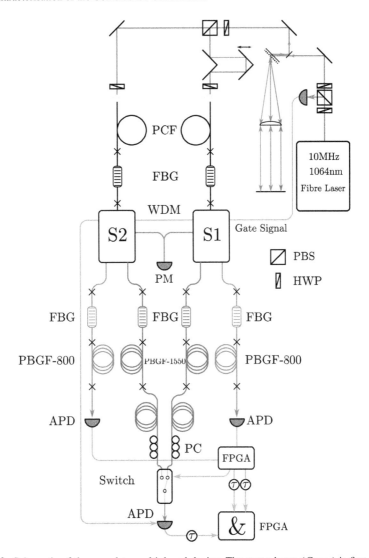

Fig. 6.2 Schematic of the complete multiplexed device. The pump beam (*Green*) is first split at a PBS to control the average power and produce a trigger signal for the InGaAs detectors. The pump then passes through a grating filter to control the bandwidth and *central* wavelength of the pulses. After this, the pump is coupled into the PCFs using aspheric lenses. The heralding APDs feed forward to the switch via the FPGA to set the state of the switch ahead of the idler photons, which are then detected by an InGaAs detector. A second FPGA is used to count coincident detector signals

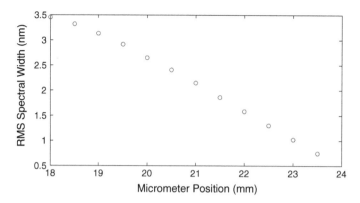

Fig. 6.3 A calibration graph for the pump filter, relating the root-mean-square bandwidth (RMS) (σ_p^{RMS}) of the signal to the translation stage position

A further HWP after the filter was used to prepare the pump polarisation, before passing through a second PBS to split the beam line into two parts, one for each source. By adjusting the position of the HWP the power was evenly distributed across each arm. The two beam paths were directed to the input of each individual source, including a variable delay line in the path to source 1. A HWP and polariser before the input to each fibre allows a particular polarisation axis of the fibre to be excited. The pump pulses were coupled into the PCF core using aspheric lenses and precision 3-axis translation stages.

Everything following was completely integrated in optical fibre. A major improvement over existing heralded single photon sources, is that this system requires no further alignment. The 1060 nm output of the WDMs were routed to an InGaAs power meter to monitor the amount of pump light and to aid coupling into the fibres. However, the pump power measured at the output port is not the actual pump power coupled into fibre, but rather the residual pump power after an FBG and the WDM. The power coupled into the PCF can be retrieved by using the applying transmission loss of −27.3 dB as a scale factor to the residual pump power. The FC-connectorised outputs of the signal arms were connected to the inputs of the silicon detectors. The common output of the switch was spliced to a in-line fibre polariser (∼1 dB insertion loss), and then sent directly to the input of the *ID210*. By adjusting the polarisation controllers on the fibre delays, in conjunction with the HWP before the PCF, any polarisation axis of the PCFs can be excited, and then translated to pass through the polariser. This ensures that we only collect idler photons in the same polarisation from both sources.

The central wavelength of the filter was set to ∼1064 nm and the bandwidth varied to optimise the count rates, finally fixing at $\sigma_p^{RMS} = 2.7$ nm. This bandwidth is contemporary with the setting used when determining the experimental JSI, and so should be well matched to the length of the fibre and therefore produce a factorable photon pair state. At this fixed bandwidth setting, the singles (N_I, N_{H_1} and N_{H_2})

6.2 Characterisation of the Coincidence Count Rates

and coincidence count rates (N_{H_1I} and N_{H_2I}) were measured as the pump power was varied, where each point was integrated for 60 s. When multiplexing, we also measure the three-fold coincidence count rate ($N_{H_1H_2I}$) between the two heralding detectors and the idler detector. When calculating the multiplexing count rate, $N_c^{(2)}$, the three-fold count rate is subtracted from the two-folds,

$$N_c^{(2)} = N_{H_1I} + N_{H_2I} - N_{H_1H_2I}. \qquad (6.7)$$

This removes those events where both sources generate a pair, ensuring that these events are not double counted. The number of accidental counts per second for the multiplexed source is calculated as,

$$N_a^{(2)} = \frac{N_{H_1} \cdot N_I}{R_p} + \frac{N_{H_2} \cdot N_I}{R_p} - \frac{N_{H_1} \cdot N_{H_2} \cdot N_I}{R_p^2}, \qquad (6.8)$$

where again the subtraction of the third term on the right hand side prevents double counting. The CAR of the 2 × 1 multiplexed source is calculated as,

$$\text{CAR}^{(2)} = \frac{N_c^{(2)} - N_a^{(2)}}{N_a^{(2)}} \qquad (6.9)$$

The pump power to each source was tailored to achieve similar coincidence count rates from each individual source.

Figure 6.4 shows the measured coincidence count rates as the pump power was varied. Source 1 requires significantly more average power to achieve the same count rate as source 2. However, the power measured here is the residual pump power after a FBG and WDM. The loss at 1064 nm in each source may be slightly different, resulting in this discrepancy. The error bars are calculated assuming Poissonian statistics for the detection events distributed over the total number of pulses in the integration time. All three sources exhibit increasing count rates with increasing power. This increase should behave quadratically with increasing pump power. However, here the behaviour appears to be linear. The dead time of InGaAs detector was set to 10 μs, the maximum potential count rate is therefore 100 kHz. The singles count rates at the detector are approximately 10% of this value indicating that we are far from saturating the possible time bins.

All of the data points plotted above were collected with the pump laser system running at 10 MHz. However, the switching window is approximately 200 ns wide. As a result of this, the maximum rate at which the switch can be set is 5 MHz. If there are two heralding signals from source 2 on consecutive pulses of the laser, the switch will already be set in the correct position to route the two corresponding photons to the output. If however there is a heralding signal from source 2, followed by a heralding signal from source 1 on the next consecutive pulse, then the switch cannot be reset in time to allow the heralded photon from the subsequent pulse through to the output. This effectively increases the amount of switch loss experienced by source 1.

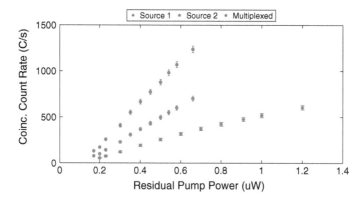

Fig. 6.4 The coincidence count rates for (*Blue*) Source 1, (*Green*) Source 2, and (*Red*) Multiplexed device. As the pump power is raised, the pair generation rate is increased and the coincidence count rate increases. Error bars are calculated assuming Poissonian statistics. By multiplexing the coincidence count rate at a fixed pump power is greatly improved

Provided that the pump power is low, and hence the probability of generating a pair per pulse is also low, then there will be very few cases where pair-generation events occur on subsequent pulses. As the pump power is increased, the number of potential heralding events from source 2 will increase. Therefore, there is more potential for missing heralded idler photons from source 1, this will eventually lead to a saturation of the relative contribution of source 1 to the total multiplexed coincidence count rate. The initial switching time window of 200 ns was chosen to ensure, in the first stages of development, that the photons arriving from source 2 would successfully be routed through. By reducing the amount of time the switch is open around the arriving heralded idler photon, this effect could be minimised. The switching window is fundamentally limited in its minimum duration by the rise and fall time of the switch, both of which require approximately 60 ns. Therefore, with the switch used here, the minimum time window that could be achieved is 120 ns. This is still too long be able to respond to pair generation from consecutive pulses. The response time of the switch is largely determined by the switch driver unit [4].

From the measured count rates we next calculate the coincidence-to-accidentals ratio using Eq. 6.6 for the individual sources and Eq. 6.9 for the multiplexed source, the results of which are shown in Fig. 6.5. For the single sources, we see that at high powers the achievable CAR is limited by noise from Raman shifted pump light, multiple pair generation events and other parasitic non-linear effects. As the pump power is reduced, the CAR increases rapidly. The CAR and coincidence count rates for each individual source were averaged to provide a measure for the average performance of a single source to which the multiplexed device can be compared, shown in cyan Fig. 6.5, with error bars of 1 standard deviation. For each of the source

6.2 Characterisation of the Coincidence Count Rates

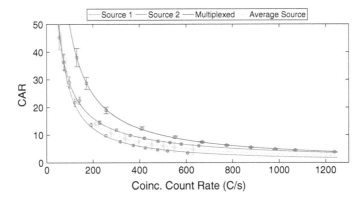

Fig. 6.5 Coincidence-to-Accidentals ratio as a function of the coincidence count rate for source 1 (*Blue*), source 2 (*Green*), and the multiplexed device (*Red*). *Curves* are fitted to the scatter points using a power law. The average source performance, calculated using the CAR and coincident count rate for sources 1 and 2 is shown (*Cyan*). By multiplexing, at a fixed CAR the heralded coincidence count rate is improved by a factor of 86%. Error bars are calculated using Poissonian statistics, except for the average source performance where 1 std. is used

Table 6.1 Coincidence-to-accidentals fitting results for curves in Fig. 6.5

	a	b	R^2
Source 1	2634	−1.007	0.9987
Source 2	1282	−0.08258	0.9997
Multiplexed	5148	−1.007	0.9997

configurations in Fig. 6.5 a curve of best fit was produced. Each set of data was fitted with a *power-law* of the general form,

$$y(x) = a \cdot x^b, \tag{6.10}$$

where the values a and b are fitting parameters and x is the coincidence count rate. Table 6.1 shows these parameters as a result of the nonlinear least squares fitting calculation using Matlab, together with the goodness-of-fit parameter R^2. All three fits display a high goodness-of-fit criterion.

As these fits do not describe a model of the single and pair generation probabilities for the count rates, we cannot use them to infer the overall probability of pair generation. We can however use them to make a comparison of the achieved count rates between the different source configurations at a fixed coincidence-to-accidentals ratio. From this, we can determine the improvement in the probability of delivering a heralded single photon from the multiplexed source with a fixed amount of noise. This is illustrated in Fig. 6.6. Due to slight differences in the construction of each photon pair source, (PCF variations, transmission loss, lengths of SMF) the CAR functions of the two individual sources are slightly different, therefore the curves do not overlap. The point at which they cross corresponds to the point at which the two

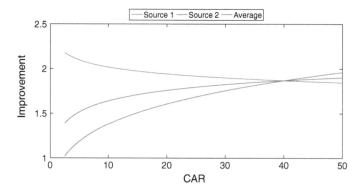

Fig. 6.6 Improvement factor at different values of CAR, calculated from the fits in Fig. 6.5, relative to source 1 (*Blue*), source 2 (*Green*), average performance of these two sources (*Red*). Due to the significant disparity in the performance of the sources, the performance relative to source 1 displays an improvement factor greater than 2. At a CAR corresponding to equal performance of the two sources, the improvement factor due to multiplexing is calculated as 1.86

sources are well matched in the CAR and coincidence count rate. At very low values of the CAR, the multiplexed source and source 2 have very similar CAR values for a given coincidence count rate, and hence the improvement factor relative to source 2 is nearly one. This is in stark contrast to source 1, where the output of the source is much poorer in terms of its CAR at a given count rate compared to the multiplexed case. In this region, the coincidence count rate of source 2 is greater than twice that of source 1. When multiplexed at this CAR, this yields an improvement factor greater than 2 relative to source 1.

Alternatively, we can choose to compare the two individual sources to the multiplexed system at the point at which they have the same CAR. This corresponds to a CAR $= 39.8$ in Fig. 6.6 and a improvement factor of 86% (1.86). This shows that we have been able to improve the probability of delivering a heralded single photon state from the output, for a fixed probability of delivering noise. Alternatively, if the heralded coincident count rate is fixed, we see an improvement in the CAR of delivered single photons corresponding to a reduction of the noise. Thus through multiplexing, we have successfully decoupled the probability of delivering a heralded single photon state from the multi-pair generation probability.

6.2.3 Experimental Results: Source 2 Noise Gating

Additionally we can also explore the use of the switch as a noise gate for a single source. When the switch is closed the output of the source is in the vacuum state, only when a heralding signal is present will the switch open to the source and route the corresponding photon to the output. We are therefore removing all those pulses

6.2 Characterisation of the Coincidence Count Rates 125

from the laser which did not result in a heralding detection, and thus most noise in the pulse train at these points [1]. We can investigate this effect by only pumping source 2, both with the switching scheme active and with it static, whilst recording the singles and coincidence count rates. For example, with the switch active, the singles count rates at the InGaAs detector are reduced, whilst leaving the coincidence counts unchanged. From these values, we again calculate the CAR and use this for the basis of our comparison.

Figure 6.7 illustrates the noise gating effect. Even at the highest average pump powers used, the CAR still has not dropped below a value of 50. This is significantly higher than any of the previous source configurations characterised above, even at the lowest pump power used. At a fixed heralded coincidence count rate, the CAR is improved by a factor of approx. 7, indicating a substantial reduction of the noise. Further improvements could potentially be made by reducing the time window that the switch is open around the pulse, removing any further noise photons that are not correlated with the pump. Figure 6.7 clearly demonstrates the added benefit of including a noise gate to this heralded photon pair source. If a noise gate was added to the output of this multiplexed source we would expect a similar level of improvement in the CAR, on top of that achieved by multiplexing. This would allow us to work in a region with higher average power and hence higher pair generation rates and thus larger coincidence count rates, whilst still minimising the noise effects.

However, this effect is only made possible due to the fact that the majority of laser pulses do not result in a pair being produced. If we were able to successfully multiplex enough sources such that the probability of finding a heralded single photon on any one pump pulse tends to 1 then the number of pulses that would be gated out tends to 0. By including a noise gate, we are not directly affecting the probability of

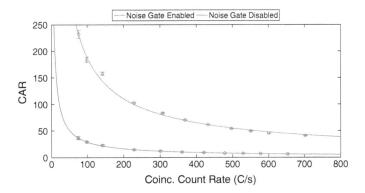

Fig. 6.7 The effect on the CAR of using the switch as a noise gate for source 2. The performance of the source with and without out noise gating active is shown in *blue*, and *green* respectively. A significant increase in the CAR is achieved at all coincidence count rates. The switch is only open to source 2 when a heralding signal is present, therefore noise photons are screened out of the mode sent to the detector, leading to a large reduction in the background and the number of accidental coincidences observed

heralding a single photon. The noise gate does let the source be pumped at a higher pump power, but this also increases the noise from multi-pair events that the noise gate cannot eliminate.

Ultimately the performance of a noise gated source is limited by the peak of the thermal probability distribution. At a mean photon number per pump pulse of $\bar{n} = 1$, this corresponds to a probability of generating pair of 0.25. Even if a detector with perfect photon number resolving detection capability were used as the heralding detector to control the switch, such that only single-pair events were routed through the gate, as well as perfect collection of the signal and idler photons, the maximum probability of heralding a single photon from the output would still only be 0.25. Only through multiplexing can this fundamental limit be exceeded and a pseudo-deterministic source of heralded single photons be produced.

6.3 Characterisation of the Second Order Coherence

One of the most fundamental tests of a heralded single photon source, is that the output does in fact consist of single photons, and that, the probability of there being a second photon present simultaneously in the mode is small. This can be done by measuring the second order coherence function, or $g^{(2)}(0)$, of the output mode.

This measurement has become commonplace in the characterisation of heralded single photon sources and is a crucial metric by which these sources can be compared [5–7].

In context of the work presented in this thesis, we have experimentally determined $g^{(2)}(0)$ of the individual heralded single photon sources and the complete multiplexed device. To do this a 50:50 fibre coupler was inserted between the output of the switch and the detector. The output modes of the coupler were sent to two InGaAs detectors. We measured the singles count rates at the heralding and idler detectors (N_{H_1}, N_{H_2}, N_{I_1}, N_{I_2}) as well as the coincidence count rates ($N_{H_1 I_1}$, $N_{H_1 I_2}$, $N_{H_2 I_1}$, $N_{H_2 I_2}$, $N_{H_1 I_1 I_2}$ and $N_{H_2 I_1 I_2}$). The experimental heralded coherence function for source j has the form [7, 8],

$$\hat{\gamma}^{(2)}_{S_j}(0) = \frac{N_{H_j I_1 I_2} \cdot N_{H_j}}{N_{H_j I_1} \cdot N_{H_j I_2}}, \qquad (6.11)$$

where we are measuring the conditional probability that both idler detectors fire given that the heralding detector has also fired.

It should be noted that the experimental coherence function, $\hat{\gamma}^{(2)}_{S_j}(0)$, does not generally equal $g^{(2)}(0)$ [8]. Only if the probability of generating multiple-pairs is small, such that $P(1) \gg P(2) \gg P(n > 2)$ does $\hat{\gamma}^{(2)}_{S_j}(0) \simeq g^{(2)}(0)$, in which case,

$$\hat{\gamma}^{(2)}_{S_j}(0) \simeq g^{(2)}(0) \simeq \frac{2P(2)}{P(1)^2}. \qquad (6.12)$$

6.3 Characterisation of the Second Order Coherence

For the multiplexed device the above set of measurements is incomplete. We must also measure $N_{H_1 H_2 I_1}$, $N_{H_1 H_2 I_2}$ and $N_{H_1 H_2 I_1 I_2}$ so that we can subtract these from the relevant count rates and remove those pulses on which both heralding detectors fire,

$$N^H = N_{H_1} + N_{H_2}, \tag{6.13}$$

$$N^{I_1}_{H_1} = N_{H_1 I_1} - N_{H_1 H_2 I_1}, \tag{6.14}$$

$$N^{I_2}_{H_1} = N_{H_1 I_2} - N_{H_1 H_2 I_2}, \tag{6.15}$$

$$N^{I_1}_{H_2} = N_{H_2 I_1} - N_{H_1 H_2 I_1}, \tag{6.16}$$

$$N^{I_2}_{H_2} = N_{H_2 I_1} - N_{H_1 H_2 I_1}, \tag{6.17}$$

$$N^C = N_{H_1 I_1 I_2} + N_{H_2 I_1 I_2} - N_{H_1 H_2 I_1 I_2}. \tag{6.18}$$

In this manner we remove the possibility of double counting these events. The experimental second order coherence function of the complete multiplexed device is then,

$$\hat{\gamma}^{(2)}_{Multi}(0) = \frac{N^C \cdot N^H}{(N^{I_1}_{H_1} + N^{I_1}_{H_2}) \cdot (N^{I_2}_{H_1} + N^{I_2}_{H_2})}. \tag{6.19}$$

In addition to the photon number statistics, the second order coherence function can also be used to determine the number of spectral modes present in the output and with it the purity of the heralded photons, see Sect. 6.3.1.

6.3.1 Experimental Results: Marginal Second Order Coherence

The *marginal* second order coherence, $g^{(2)}_m(0)$, reflects the fact that we are only measuring one arm of the source, ignoring all heralding signals from the other arm. In doing so, we are measuring the idler probability distribution whilst averaging over the signal probability distribution, as illustrated in Fig. 6.8. The goal of the phasematching scheme we have implemented in each source is to produce photon pairs into a single joint spectral mode. By performing this measurement we can experimentally determine the number of spectral modes in which the idler photon is generated, and therefore determine how successful we have been in heralding a pure single photon state.

Using the values of K extracted from the marginal second order coherence, we can compare to the calculated number of spectral modes extracted from the stimulated emission tomography plots of the JSI. As we have seen in Sect. 4.3, stimulated emission tomography allows us to measure a joint spectral distribution function that is proportional to the JSI, where the constant of proportionality is the average number of photons in the seed beam.

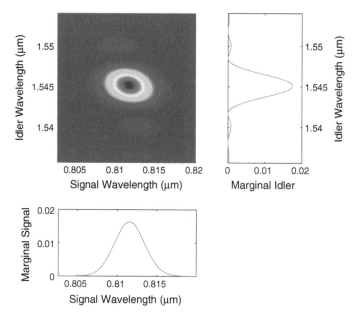

Fig. 6.8 Simulated JSI and the marginal distributions of the signal and idler photons. By measuring the marginal distribution of the idler photons, the signal photon probability distribution is averaged over. If the JSI is factorable then, each marginal distribution is composed of only one spectral function. Detection of one photon will not affect the marginal distribution of the remaining photon

To perform the marginal second order coherence measurement, a 50:50 fibre coupler is added to the output of the switch, as seen in Fig. 6.9. This allows us determine with what probability the two idler detectors register a coincidence simultaneously and with it the photon number statistics of the idler arm.

The outputs of the 50:50 coupler were sent to the existing *ID210* and a further *ID201* InGaAs detector. The detector triggers and timing delays were scanned to find the peak of maximum coincidence counts with the heralding signals of each source. In doing so, the detection windows of the two detectors become synchronised in time. At the FPGA-correlator, the singles count rates at each idler detector (N_{I_1} and N_{I_2}) and the number of coincidences between the two idler detectors ($N_{I_1 I_2}$) were measured. The remaining degrees of freedom were then used to optimise the JSA and $g_m^{(2)}(0)$. As the length and dispersion of the PCF are fixed, the only thing left available to change is the pump configuration. The two easily adjustable pump parameters are the bandwidth and central wavelength of the laser pulses. Other factors such as the chirp on the pump pulses entering the fibre and the amount of SPM acquired on propagation in the fibre are harder to control.

Here, we choose to fix the central wavelength, and vary the spectral bandwidth of the pump pulses, a calibration curve of the separation of the filter blades against the root mean square bandwidth was produced and is shown in Fig. 6.3. As the filter bandwidth is cut down, the pulses broaden in the temporal domain resulting in a

6.3 Characterisation of the Second Order Coherence

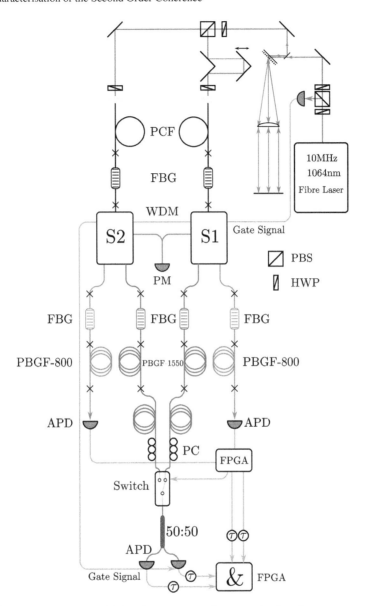

Fig. 6.9 A schematic of the multiplexed source, set-up for a second order coherence measurement. The set-up is identical to Fig. 6.2, with the exception of the inclusion of a 50:50 coupler on the output of the switch and a second InGaAs detector. For the marginal measurement: The switch is no longer set by the heralding signal and is instead statically coupled to either source 1 or 2. The coincidences between the two InGaAs detectors are then measured independently of any heralding detector

reduction in the peak power of the laser pulses. As a result of this the amount of SPM acquired on propagation will vary with bandwidth.

The coincidence count rates were measured for Source 1, Source 2 and the multiplexed device, for a range of bandwidths at a fixed average power. Due to low detection efficiencies, the two-fold coincidence count rate between the idler detectors was low, necessitating a long integration time of 15 mins. The experimental marginal coherence function has the same form as the heralded coherence function in Eq. 6.11, with the exception that the simultaneous detection of two idler photons is no longer conditional on whether there was a heralding event. Instead, it is dependent on whether or not there is a pump pulse present, and so the heralding count rate N_{H_j} is replaced with the repetition rate R_p,

$$\gamma^{(2)}_{marg.}(0) = \frac{N_{I_1 I_2} \cdot R_p}{N_{I_1} \cdot N_{I_2}} \simeq g_m^2(0). \tag{6.20}$$

Equation 6.20 is plotted as a function of the RMS bandwidth in Fig. 6.10a, b for source 1 and source 2 respectively, with no background subtracted.

Initially, with the pump bandwidth set to the narrowest position, $g_m^{(2)}(0)$ is low as the pump envelope function is not well matched to that of the fibres' phasematching function, this results in an anti-correlated JSI, see left hand panel of Fig. 6.11. As the bandwidth is increased the marginal second order coherence rises as the pump envelope and phasematching functions become matched producing a factorable JSI, see centre panel of Fig. 6.11. Finally as the bandwidth is increased further, the JSI broadens parallel to the signal axis, but remains factorable as the pump and signal are asymmetrically group-velocity matched, see right hand panel of Fig. 6.11. As the bandwidth increases the pulse duration decreases and the peak power of the pulse rises at a fixed average power. As a result of this the pulse acquires more SPM as it propagate through the PCF and breaks up. This leads to a reduction in $g_m^{(2)}(0)$ at large bandwidths. For source 2 this behaviour is clearer to see, as $g_m^{(2)}(0) \sim 1.4$ at a RMS bandwidth of 0.7 nm, before rising gradually and reaching a peak of $g_m^{(2)}(0) \sim 1.7$ at $\sigma_p^{RMS} \sim 2.5$ nm, after which it begins to reduce. The results for source 1 are less clear, but a rise is still observed with a peak value of $g_m^{(2)}(0) \sim 1.7$ at $\sigma_p^{RMS} \sim 1.5$ nm.

From the peak value of $g_m^{(2)}(0)$, the minimum value of $K = 1.42$ was calculated for both source 1 and source 2. This indicates the presence of less than two Schmidt modes in each source. Through the use of Eq. 2.32, we calculate the purity of the reduced idler state as 70% for both source 1 and source 2. We can now compare the K parameter values extracted from the marginal second order coherence measurement, with those calculated from the stimulated JSIs. The stimulated emission technique for measuring the JSI does not yield any phase information and so we cannot reconstruct the JSA, on which the Schmidt mode decomposition should be carried out. We can approximate the JSA by taking the square root of the JSI, and then apply the singular value decomposition to this matrix, to determine the number of Schmidt modes present. Therefore, we can only use this to place a lower (upper) bound on the Schmidt-mode number (Purity). Using the singular value decomposition technique

6.3 Characterisation of the Second Order Coherence

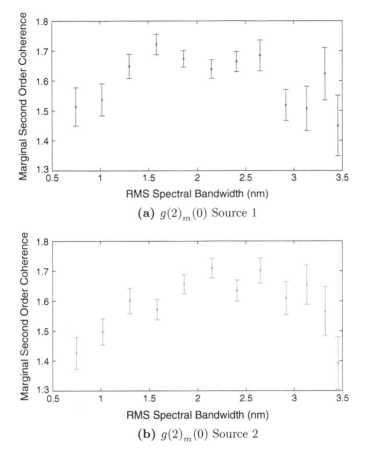

Fig. 6.10 Marginal second order coherence with varying pump bandwidth, *top* Source 1, *bottom* Source 2. Both sources exhibit a similar behaviour producing a factorable state with a purity of 70% over a broad range of bandwidths

on the approximate JSA [10], the Schmidt number, was found to be $K = 1.16$ and $K = 1.15$, corresponding to a reduced idler purity of 86.2 and 86.9% for source 1 and source 2 respectively.

The numerical simulations conducted during the PCF design phase predicted a potential heralded state purity of 94%. The experimentally determined values for the purity of the state are less than this. However, the simulation conditions are for a set of idealised circumstances, including a perfectly homogeneous PCF cladding structure (in both transverse and longitudinal directions) producing perfect phasematching, whose length can be chosen to perfectly match the transform limited pump pulses with a Gaussian spectral profile. Considering these effects, we have extracted a high level of experimental performance from the sources.

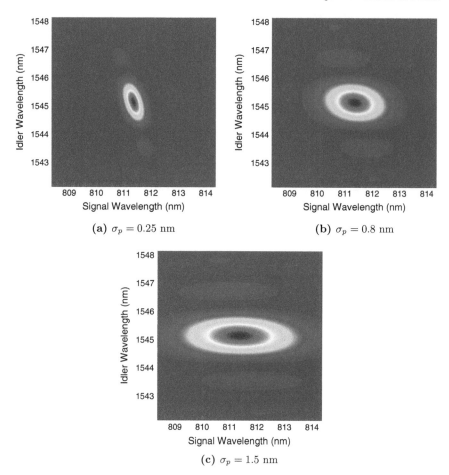

Fig. 6.11 Simulated JSIs for L = 0.25 m, using different pump bandwidth parameters: **a** $\sigma_p = 0.25$ nm, **b** $\sigma_p = 0.8$ nm, and **c** $\sigma_p = 1.5$ nm. At narrow bandwidths the JSI develops some anti-correlation (*left*). As the pump bandwidth is increased it becomes matched to the fibre phasematching functions, producing a factorable state. Due to the asymmetric group-velocity matching scheme implemented in the fibre, the state remains factorable over a large range of bandwidths. However, eventually the JSI will become *curved*, due to higher-order dispersion, and become correlated [9]

The marginal second order coherence of the sources could possibly be improved with further work on the control of the pump pulses and detection electronics. The fibre laser used in the experiments produces pulses with a large amount of noise of the frequency spectrum, most likely due to SPM in the amplifier, which is nearly unavoidable in a fibre laser. It was shown by Bell et al. [11] that the SPM acquired by the pump pulses on propagation in the pair generation medium, has a detrimental effect on the purity of the heralded state. The pulses used in this experiment already possess some amount of SPM before entering the fibre. This is compounded in the fibre where the extra SPM that is acquired will also be large due to the large nonlinearity of the fibre, as well as the high peak power of the laser. Moving to a laser system using a free space cavity, such as a Ti:Sapphire, would remove the possibility of any SPM on the initial pump pulses, potentially allowing us to achieve a much higher heralded state purity. In addition to this, the spectral profile of the laser pulses used is far from the assumed Gaussian profile, possessing sharp edges with a square profile. This is a result of the hard edges of the filter blades in the filter. A further benefit of the free space cavity laser is the smooth spectral-profile pulses.

To the best of our knowledge, this is the first implementation of a multiplexing scheme where the generated photons have been spectrally engineered, and the first experimental demonstration of high levels of purity from photons generated in a multiplexed device.

6.3.2 Experimental Results: Heralded Second Order Coherence

Following on from the measurement of the marginal second order coherence function, the heralded second order coherence, $g_H^{(2)}(0)$ was investigated. For a perfect single photon source we expect $g_H^{(2)}(0) = 0$, where a value of $g_H^{(2)}(0) < 1$ is indicative of the preparation of non-classical states of light. The pump filter bandwidth was tuned to $\sigma_p^{RMS} = 2.65$ nm corresponding to the peak of the measured marginal second order coherence functions.

The heralded experimental coherence functions for each individual source and the multiplexed system were measured as the pump power was varied. From the raw count rates the coherence functions of Eqs. 6.11 and 6.19 were evaluated. These are plotted against the measured coincidence count rate in Fig. 6.12. Once again the disparity between source 1 and 2 is clear to see. There is evidently more noise in source 1, at a fixed coincidence count rate, due to uncorrelated photons and multi-pair emission. Overall, all three systems display the same behaviour. At low pump powers, the second order coherence is low. As the power is raised, the number of coincidence counts increases, but so does $g_H^{(2)}(0)$, as more noise photons are generated.

The approximately linear increase in $g_H^{(2)}(0)$ can be mainly attributed to photons generated into the long-wavelength arm from spontaneous Raman scattering and also multi-pair FWM events. Ultimately it is these two contributions that limit the achievable value of $g_H^{(2)}(0)$ at high powers. At low pump powers, the relative contribution to the raw count rates from APD dark counts is large, this limits the achievable heralded second order coherence in this region. For a fixed coincidence count rate, we see that by multiplexing we have been able to reduce $g_H^{(2)}(0)$. This corresponds to a suppression of the noise from higher-order photon pairs. Alternatively, at a fixed $g_H^{(2)}(0)$, we can achieve an increased probability of delivering a single photon.

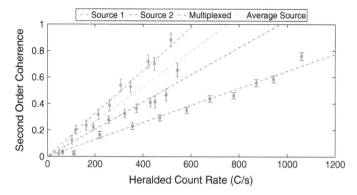

Fig. 6.12 Heralded second order coherence function, $g_H^{(2)}(0)$, increases linearly with coincidence count rate, due to the increase in noise photons from spontaneous Raman scattering and multi-pair generation events: (*Blue*) Source 1, (*Green*) Source 2, (*Red*), multiplexed device, (*Cyan*) average individual source performance. Error bars are calculated using Poissonian statistics

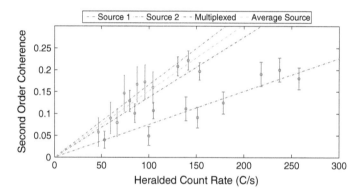

Fig. 6.13 Heralded second order coherence function, $g_H^{(2)}(0)$, sampled at lower coincidence count rates: (*Blue*) Source 1, (*Green*) Source 2, (*Red*), multiplexed device, (*Cyan*) average individual source performance. At a fixed $g_H^{(2)}(0)$, the count rate is improved by multiplexing for a fixed amount of noise. The minimum value of $g_H^{(2)}(0) = 0.049$ indicates the multiplexed device is capable of producing a high quality single photon output state

To provide the best quality single photon state we must operate in a region where $g_H^{(2)}(0) < 0.1$. This region of the plot was re-measured over a larger number of samples and is displayed in Fig. 6.13. The same overall behaviour, as seen in the broad scan, can be seen here. In Fig. 6.13 the lowest heralded second order coherence was recorded as $g_H^{(2)}(0) = 0.049 \pm 0.02$ at a coincidence count rate of $100 \pm 10 C/s$, indicating that this is a high performance source in terms of the single photon output.

6.4 Summary of Source Performance and Potential Improvements

From the heralded count rates we can determine the pair generation probabilities of each source and the complete multiplexed device. At a CAR = 39.8, we have achieved a heralded count rate of 67 ± 8 C/s, 67 ± 8 C/s and 124 ± 11 C/s for source 1, source 2, and the multiplexed device. At a repetition rate of 10 MHz this corresponds to a pair generation probability of 6.7×10^{-6}, 6.7×10^{-6} and 1.25×10^{-5} respectively. Whilst these values are low, the benefit of multiplexing is very clearly seen.

In order to improve these count rates and corresponding pair generation probabilities the sources should be operated at higher average powers. However in the current implementation the sources, the achievable count rate is limited by the need to maintain a high CAR which is degraded at higher powers by the presence of noise photons reaching the detectors. This is particularly noticeable in the signal arm of each source where the rate of detection events rapidly increases as the pump power is raised without a commensurate increase in idler photon detections. A clear indication of noise reaching the detector. The noise likely has a number of origins. Firstly, more pump photons breaking through the filtering and reaching the silicon APDs which are still able to detect the photons at the pump wavelength. Secondly, on inspection of the transmission bands of the PBGF filters and WDM, Stokes shifted pump light due to Raman scattering will propagate in the signal arm and again reach the APDs where it can be detected. Finally, as a result of the phasematching scheme implemented in the PCFs, photon pairs are also generated on the inner side bands of the phasematching contours. Four-wave mixing in this region is present in the FWM spectrum of Fig. 4.7b, and will again be guided in the signal arm of the source through the PBGFs albeit with high attenuation. All of these processes contribute to an increase in the rate of spurious heralding events at the silicon APD without a corresponding idler photon being detected. This significantly reduces the CAR and heralding efficiency of the sources, and in turn limiting the power level at which they can be pumped.

Another source of accidental coincidence count rate are those generation events that result in multi-pair emission, which increases with pump power and is a fundamental limitation in photon-pair sources. At present, the heralding detectors have no ability to discriminate single pair generation from multi-pair generation. As a result of this multi-pair events can be heralded as single-pair events leading to an increase in the number of accidental coincidences. As we have seen in Chap. 3, moving from a binary detector to a pseudo-PNR or true PNR detector gives a significant performance enhancement of the individual sources as they can be operated at higher individual pair pulse probabilities.

Further to spectrally filtering the output of the photon-pair sources to remove uncorrelated noise photons from the output, we have also demonstrated the potential of temporal filtering in the form of a noise gate. Because photon number is correlated between the arms, the heralding signal can be used to effectively induce a time dependent loss in the heralded arm, such that only photons in the idler arm that have a correlated heralding signal can propagate through to the output. We have seen in Sect. 6.2.3 that by including a noise gate we can improve the CAR of the individual source by a factor of 7. This has the potential to be improved further by reducing the time window that switch opens around the arriving heralded photon and also by combining it with detector that has some PNR capabilities. The noise gate therefore not only acts as a filter for uncorrelated photons but also as a filter for multi-photon events. Together this could allow each individual source to be operated closer to the peak of the thermal photon number probability distribution, which when multiplexed yields a significant improvement in the performance of the device. This also has the added benefit of reducing the number of individual sources that need to be multiplexed together to produce a pseudo-deterministic single photon source. We have implemented the noise gate by using the optical switch that was used for multiplexing, however if we wish to also multiplex the two sources a second switch or fibre coupled optical modulator would be required. The added benefits of a fibre coupled optical modulator are the low loss and high switching speeds (upto GHz bandwidth) that can be achieved.

One of the defining components of the multiplexed set-up is the switch. The switch currently in use has a reasonable efficiency of 80%, but only has a switching bandwidth of 1 MHz. As the pump laser system operates at 10 MHz, the switch can not react to every pulse from the laser, this is manifested as an increase in the insertion loss of the switch, which as we have seen in Chap. 3 must be minimised. The main limitation in the switching speed is the switch driver electronics. A new design of switch driver could allow the switch to operate faster and therefore be able to respond to every possible heralding signal. Ultimately, the switching speed will be limited by the performance of the material. Developing an ultra fast switch with up to several gigahertz of bandwidth would allow access to ultra fast laser systems such as the Ti:Sapph and VECSEL, increasing the base rate at which heralded photons can be delivered from the multiplexed source.

With the current level of switch technology, a further two levels of switching network could be installed before the induced loss outstrips the benefit of multiplexing. This would correspond to a total of 8 individual sources multiplexed together and a maximum improvement factor of 8. This would require a large number of components and identical pieces of PCF, so alternatively the already spatial multiplexed sources could be combined with a delay loop to produce a hybrid temporal-spatial scheme (2 spatial bins and 4 temporal bins). This scheme combined with pseudo-PNR controlled noise gating would represent a significant step forward in the performance of the multiplexed system constructed as part of this thesis.

In addition to spurious counts from the signal arm, the overall coincidence count rate is also determined by the loss leading to the idler detector and the efficiency of the detector itself. In these characterisation measurements the heralding detector is an important component, and as we have seen in Chap. 3 optimising its efficiency is key to overall source performance. Unlike the heralding detector, the idler detector

6.4 Summary of Source Performance and Potential Improvements

is not a component of the source and is only present as a tool for characterisation. In all of the above experiments the InGaAs detector was run in a gated configuration with a detector efficiency of 15%. This efficiency could be corrected for to give the rate of delivered heralded single photons, increasing the overall heralded rate by a factor of approximately 6.

Another factor affecting the pair generation probability per pulse is the length of the PCF used in each source. The probability of generating a pair is proportional to the square of the fibre length. Thus by increasing the length of PCF used we could increase rate of photon-pair generation. However, as was seen in Chap. 4 the PCF is not uniform over length scales greater than 30 cm. As we also wish to deliver a spectrally pure state these variations in the PMF and JSA must be minimised and we are therefore limited in the length of fibre which can be used. Improving the fabrication technique to reduce the degree of variation in the PCF structure along its length would allow a longer length of fibre to be used and with it the potential of higher pair-generation probabilities.

One key requirement of the heralded single photons from the multiplexed source is that they must be completely indistinguishable from one another. In Sect. 4.4.2 we saw that the stimulated joint spectra of the two sources shows an overlap of 95% indicating a high degree of spectral indistinguishability. This measurement does however neglect the remaining degrees of freedom of the photon such as polarisation state and the timing. A more suitable approach is to perform HOM interferometry on the heralded photons from the two sources. This also provides an indication of the suitability of these photons to be used in quantum information schemes that are based on HOM interference. The experimental evidence of indistinguishability in a HOM interference experiment is the drop in coincidence count rate between two detectors placed in the output ports of the beam splitter as some distinguishing information (such as time of arrival at the beam splitter) is introduced to one of the photons. This is know as the HOM dip. The depth, or visibility, of which indicates the indistinguishability of the two photons impinging on the beam splitter. For perfectly indistinguishable photons the count rate drops to zero and visibility tends towards 1.

However the visibility of the dip is not just dependent on the indistinguishability of the two photons but also on the purity of each single photon state. Overall the visibility, V is given by,

$$V = \frac{\mathcal{P}_1 + \mathcal{P}_2 - \mathcal{O}(\hat{\rho}_1, \hat{\rho}_2)}{2}, \qquad (6.21)$$

where $\mathcal{P}_1$, $\mathcal{P}_2$ are the purities of the heralded single photons from source 1 and 2 respectively and $\mathcal{O}$ is the distinguishability of the density matrices of source 1 and 2. For perfectly indistinguishable photons $\mathcal{O} = 0$, but V will only attain its maximal value of 1 if both $\mathcal{P}_1 = \mathcal{P}_2 = 1$. Using Eq. 6.21 we can infer the potential HOM dip visibility of the two sources from the spectral overlap of 95% and the Schmidt mode number $K = 1.42$. Assuming that remaining degrees of freedom of the photon can be engineered into a single mode (e.g. using polarisers), and the two photons are perfectly aligned in time so that they impinge on the beam splitter at precisely the

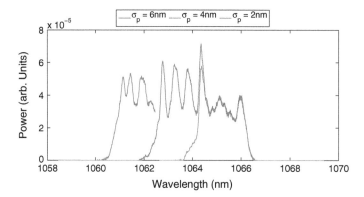

Fig. 6.14 Pump spectra after filtering in the 4-f spectrometer. The pulse spectrum exhibits very sharp edges due to the aggressive filter edges and a significant amount of noise on the *top* of the spectrum, although some of this could be due to modal interference in the multimode fibre used in the collection optics

same time, the distinguishability of the photons is determined mostly by the spectral overlap which is very high. Therefore, an upper bound of HOM dip visibility is determined mostly by the average purity of the heralded photons.

A direct measure of the purity of the heralded photons from source 1 and 2 was made in Sect. 6.3.1 by determining the marginal second order coherence of the idler photons generated in each source. This yielded a Schmidt mode number of $K = 1.42$ for both sources equivalent to a purity of $\mathcal{P}_1 = \mathcal{P}_2 = 0.7$. Substituting these values into Eq. 6.21 gives a HOM dip visibility of $V = 0.7$. The purity of the idler photons could be improved, and with it the HOM visibility, by improving the conditioning of the pump pulses to produce a pulse shape that more closely resembles the ideal gaussian distribution. The pump envelope spectrum is shown in Fig. 6.14. By passing the pump pulses through a 4f-spectrometer the bandwidth and central wavelength of the pulses was controlled, but the resulting pulse spectra is no longer the ideal Gaussian profile that is assumed for many calculations. Additionally, because the pulses are produced from an amplified fibre laser the pulse spectrum shows characteristics of SPM. This leads to chirping of the pump pulse and because of it additional phase factors in the JSA, something which the SET process that was used to determined the two-photon correlation is insensitive to. Reducing the uncorrelated photon noise and moving to a different laser system, or establishing a method of dispersion compensation to remove the chirp on the pump pulses, could lead to a significant improvement in the marginal second order coherence ($g_m^{(2)}(0) \to 2$ and $\mathcal{P} \to 1$) and hence the purity of the idler photons and HOM dip visibility.

6.5 Conclusions

This chapter has described the implementation and characterisation of a spatial multiplexing scheme for the two individual photon pair sources constructed in Chap. 5. The two sources were integrated into one device through the use of a fast electro-optic switch, where the heralding detector of source 2 controls the state of the switch in a feed-forward manner.

The individual sources and multiplexed device were characterised first in terms of the coincidence-to-accidental ratio. A CAR in the range of 5-50 was achieved depending on the coincidence count rate. At a CAR = 10, coincidence count rates of 245 ± 15C/s, 356 ± 18C/s and 491 ± 22C/s were achieved for source 1, source 2, and the multiplexed device when pumped at repetition rate of 10 MHz. For a fixed CAR = 39.8, an improvement of 86% was achieved in the heralded single photon rate by multiplexing two sources. This corresponds to an increase in the probability of delivering a single photon at a fixed level of noise.

Following these measurements, the marginal second order correlation was measured for each source. Through this the purity of the heralded state can be determined. The bandwidth of the pump pulses was varied and the purity maximised. For both sources, the peak value of the marginal second order coherence function was found to $g_m^{(2)}(0) = 1.7$. This corresponds to a Schmidt number of $K = 1.42$ and a heralded state purity of 70%.

Finally, the output of both individual sources and the multiplexed system was shown to be non-classical in nature by violating the Cauchy inequality, indicating that all configurations of the sources were generating heralded single photon states. The measured values of $g_H^{(2)}(0)$ here are comparable to other multiplexed sources based on both PDC and FWM [5, 12–14]. In some cases surpassing the heralded count rates of previous multiplexed devices by up to an order of magnitude, whilst achieving a consistent $g_H^{(2)}(0)$[12, 13]. Through better spectral conditioning of the pump pulses and reducing optical losses this performance could be extended further.

Although the absolute performance of the individual sources is not, in terms of the heralded count rates, superior to other sources using FWM or PDC, the multiplexing scheme that has been implemented here could be applied to many of these sources. In doing so, a comparable improvement in the heralded photon count rate at a fixed signal-to-noise ($g_H^{(2)}(0)$ or CAR) could be achieved.

References

1. G. Brida, I.P. Degiovanni, M. Genovese, A. Migdall, F. Piacentini, S.V. Polyakov, I.R. Berchera, Experimental realization of a low-noise heralded single-photon source. Opt. Express **19**, 1484–1492 (2011)
2. C. Eckart, F.R. Shonka, Accidental coincidences in counter circuits. Phys. Rev. **53**, 752–756 (1938)

3. A.R. McMillan, J. Fulconis, M. Halder, C. Xiong, J.G. Rarity, W.J. Wadsworth, Narrowband high-fidelity all-fibre source of heralded single photons at 1570 nm. Opt. Express **17**, 6156–6165 (2009)
4. H. Jiang, Y.K. Zou, Q. Chen, K.K. Li, R. Zhang, Y. Wang, Transparent electro-optic ceramics and devices, Technical Report, Boston Applied Technologies Incorporated (2004)
5. X.-S. Ma, S. Zotter, J. Kofler, T. Jennewein, A. Zeilinger, Experimental generation of single photons via active multiplexing. Phys. Rev. A **83**, 043814 (2011)
6. G. Harder, V. Ansari, B. Brecht, T. Dirmeier, C. Marquardt, C. Silberhorn, An optimized photon pair source for quantum circuits. Opt. Express **21**, 13975–13985 (2013)
7. J.B. Spring, P.S. Salter, B.J. Metcalf, P.C. Humphreys, M. Moore, N. Thomas-Peter, M. Barbieri, X.-M. Jin, N.K. Langford, W.S. Kolthammer, M.J. Booth, I.A. Walmsley, On-chip low loss heralded source of pure single photons. Opt. Express **21**, 13522–13532 (2013)
8. A. Migdall, S.G. Polyakov, J. Fan, J.C. Beinfang (eds.), Single-photon generation and detection, volume 45 of experimental methods in the physical sciences (Elsevier, 2013)
9. K. Garay-Palmett, A.B. U'Ren, R. Rangel-Rojo, Tailored photon-pair sources based on inner-loop phasematching in fiber-based spontaneous four-wave mixing. Rev. Mex. de Física **57**, 15–22 (2011)
10. C.K. Law, I.A. Walmsley, J.H. Eberly, Continuous frequency entanglement: effective finite hilbert space and entropy control. Phys. Rev. Lett. **84**, 5304–5307 (2000)
11. B. Bell, A. McMillan, W. McCutcheon, J. Rarity, On the effects of self- and cross-phase modulation on photon purity for four-wave mixing photon-pair sources (2015). ArXiv e-prints
12. M.J. Collins, C. Xiong, I.H. Rey, T.D. Vo, J. He, S. Shahnia, C. Reardon, T.F. Krauss, M.J. Steel, A.S. Clark, B.J. Eggleton, Integrated spatial multiplexing of heralded single-photon sources. Nat Commun. **4** (2013)
13. T. Meany, L.A. Ngah, M.J. Collins, A.S. Clark, R.J. Williams, B.J. Eggleton, M.J. Steel, M.J. Withford, O. Alibart, S. Tanzilli, Hybrid photonic circuit for multiplexed heralded single photons. Laser Photon. Rev. **8**, L42–L46 (2014)
14. G.J. Mendoza, R. Santagati, J. Munns, E. Hemsley, M. Piekarek, E. Martín-López, G.D. Marshall, D. Bonneau, M.G. Thompson, J.L. O'Brien, Active temporal and spatial multiplexing of photons. Optica **3**, 127–132 (2016)

Chapter 7
Conclusion

It was first proposed in 2002 by Migdall et al. [1] that a pseudo-deterministic source of heralded single photons could be realised by active multiplexing. In the intervening years, this work has been built upon with the development of several other novel multiplexing protocols in a variety of domains. Of these protocols, a limited number of multiplexed devices have been constructed, but none so far have achieved this with the heralded pure single photon states required by many quantum information science applications. The work presented in this thesis builds upon a broad foundation of previous work in a range of fields, including photon-pair generation in photonic crystal fibres, spectral engineering of the two-photon state and multiplexing strategies, to demonstrate an active multiplexed source of heralded single photons in pure spectral states. How this source fits into the landscape of multiplexed sources can be seen in Table 7.1, where it is highlighted in blue. Overall, the performance of the multiplexed source is comparable to previous sources, outperforming many systems of similar multiplexing depth whilst doing so in an integrated architecture that is stable, scalable, portable, and user-friendly.

7.1 Summary

In Chap. 1, the motivation for developing a deterministic source of pure indistinguishable heralded single photons was discussed. The properties required of these photons and the demands this places on source technology were outlined. Chapter 2 developed the theory of photon pair generation via four-wave mixing in photonic crystal fibres, introducing the concept of phasematching through which photons in the pair may become spectrally correlated. Following this, a method of spectrally engineering the two-photon state to remove these correlations was detailed. This was

Table 7.1 Multiplexing scheme performance comparison. Medium = pair generation process and nonlinear material. Multiplexed = implemented multiplexing scheme (S = spatial, T = temporal). Depth = number of multiplexed bins, or stages used. Pure State = spectral engineering carried out. $g^{(2)}(0)$ = second order coherence measurement. C.R. = count rate or brightness. Rep. rate = maximum switching rate. The system outlined in this Thesis is shown highlighted in blue for comparison

Year	Paper	Medium	Multiplexed	Depth	Purity	$g^{(2)}(0)$	C.R.	Rep. Rate
2002	Pittman [2]	PDC in BBO	Storage Loop (T)	5	–	–	2 C/s	75 MHz
2004	Jeffery [3]	PDC in BBO	Storage Loop (T)	–	–	–	–	10–50 kHz
2011	Ma [4]	PDC in BBO	Log-Tree (S)	4	–	0.08	714 C/s	~15 MHz
2011	Broome [5]	PDC in BBO	Pulse Doubling (T)	2	–	–	40 C/s/mW	152 MHz
2013	Collins [6]	FWM in PhCW	Log-Tree (S)	2	–	~0.19	1–5 C/s	1 MHz
2014	Meany [7]	PDC in PPLNW	Log-Tree (S)	4	–	–	60–70 C/s	1 MHz
2015	Kaneda [8]	PDC in BiBO	Storage Loop (T)	35	0.05	0.479	19.3 kC/s	50 kHz
2016	Mendoza [9]	PDC in PPLN	Tree + Delay (S and T)	2 and 4	–	–	140 C/s	500 KHz
2016	Xiong [10]	FWM in Nanowire	Delay Lines (T)	4	69–91%	–	600 C/s	10 MHz
2016	This Thesis	FWM in PCF	Log-Tree (S)	2	70–86%	0.05	100 C/s	1 MHz

achieved by group-velocity matching the pump, signal, and idler field in a method laid out by Grice and Walmsley [11] and Garay-Palmett et al. [12]. The Schmidt decomposition technique was introduced as a means of quantifying the degree of spectral correlations within the two-photon joint spectral amplitude. From this the approximate purity of the heralded single photon was determined using the singular value decomposition. Factors affecting the photon number statistics of the heralded single photon were discussed, and the second order coherence function was introduced as a method for measuring the photon number statistics.

In Chap. 3, the theory behind active multiplexing in the spatial and temporal domain was introduced. The performance of these multiplexed sources is largely influenced by the loss of the constituent components. To determine the effect of loss, a numerical model of the photon pair generation process and multiplexing strategy was established. Within this model, systems consisting of different numbers of photon pair sources were considered with different levels of loss. It was found that in the spatial domain, when the loss of the switch is high, it is no longer beneficial

7.1 Summary

to increase the multiplexing depth, even though the probability of heralding a pair increases. Increasing the multiplexing depth becomes detrimental, resulting in a reduction in the probability of delivering a single photon from the output, due to a reduction in the fidelity of the heralded single photon state by reintroduction of the vacuum. In the temporal domain, there is no such detrimental effect by "over-multiplexing", instead the performance of the source becomes dominated by only a handful of temporal modes which have experienced the least amount of loss. A comparison between temporal and spatial multiplexing was made, from which it was seen that temporal multiplexing is more efficient in terms of the number of experimental resources required.

Chapter 4 documented the design, fabrication, and characterisation of photonic crystal fibres for photon pair generation and more specifically photons heralded into a spectrally engineered pure state. The technique of stimulated emission tomography was introduced as a means of measuring the joint spectral distribution function of the two-photon state. From these measurements, an upper bound was placed on the purity of the heralded idler photon of $\sim 86\%$. Stimulated emission tomography was also used to estimate the length scale over which the PCF yields homogeneous regions of phasematching, this was found to be roughly ~ 30 cm. Lengths longer that this typically contained multiple regions of different dispersion, resulting in fluctuations in the FWM wavelengths and additional spectral correlations along the length of the fibre.

Chapter 5 details the construction of two photon pair sources in an integrated fibre architecture, using two pieces of PCF which were identified as having factorable joint spectral distributions from the SET measurements in Chap. 4. A novel, broadband spectral filter was developed in a photonic bandgap fibre. This allowed the amount of noise in the system to be reduced without impacting on the spectrally engineered JSI, unlike previous implementations of fibre based sources which made use of narrowband FBGs. Each source was characterised in the classical domain, including measurements of the loss of 5.6 dB at 810 nm and 5.0 dB at 1550 nm, not including the switch loss of approximately 1 dB. The process of switch integration to multiplex the two sources was outlined.

In Chap. 6, each individual source and the complete multiplexed device was characterised. This included measurement of the coincidence count rates between the signal and idler modes. From this, at a coincidence to accidentals ratio of 39.8, an improvement factor of 86% was found on multiplexing the two sources together. Following this, the marginal second order coherence was measured to determine the spectral purity of the heralded single photon state. This allowed a comparison with the SET measurements of Chap. 4. Both sources displayed a $g_m^{(2)}(0) = 1.7$, corresponding to a heralded state purity of 70%.

Finally a violation of the classical bound of the heralded second-order coherence was observed for both individual sources and the complete multiplexed device. This indicated that single photon states were indeed being generated by the systems. A minimum value of $g_H^{(2)}(0) = 0.049 \pm 0.02$ at a coincidence count rate of 100 ± 10

C/s was recorded for the multiplexed device. Unlike previous systems, this multiplexed device is capable of producing high quality single photon states at count rates similar or greater than previous implementations. This demonstrates a step forward in multiplexed source technology.

7.2 Future Outlook

7.2.1 Photon Pair Generation in Fibres

The photon pair sources demonstrated in this thesis have several novel features that could yield further performance enhancement if optimised further. Firstly, the phasematching scheme implemented to give group-velocity matched solutions far from the pump was selected to work with a relatively inexpensive laser system. The output pulses from the laser have a significant amount of SPM on them; this will limit the spectral and temporal purity of the heralded state. The fibre structure that yields this phasematching is quite sensitive to imperfections in the fabrication process. Improving the fibre drawing process, particularly to reduce the sensitivity to small changes in the structure, would be beneficial. This could be achieved by working at a higher draw tension to minimise these fluctuations. If this were successful then the fibre would be homogeneous over a longer length, allowing longer pieces of fibre to be used. As the probability of generating a pair scales with the square of fibre length, this would lead to higher count rates.

Secondly, the use of photonic bandgap fibre as a broadband filter was demonstrated to be successful, but at the expense of approximately 3dB of loss per filter. Further work on minimising the insertion loss of this component, as well as reducing the bend loss, which currently limits the footprint of the source to 80 cm × 45 cm × 8 cm, would be of significant benefit. A photograph of the two complete sources is shown in Fig. 7.1. An alternative, and perhaps more elegant route would be to try to achieve photon pair generation in a hybrid photonic crystal-bandgap fibre. Several groups [13, 14] have implemented a hybrid fibre structure whose cross-sectional profile consists of an array of high-index rods to achieve a photonic bandgap, whilst a secondary array of air holes in the interstitial sites could be used to tune the dispersion. These air holes also reduce the bend loss of the fibre by increasing the refractive index contrast between the guided mode of the core and the cladding supermodes [15]. This would in theory allow photon pair generation with very little Raman noise. This would solve one of the largest deficits of fibre based photon pair sources.

The bandgap portion of the cladding leads to the formation of regions of high and low transmission as seen in Chap. 4. Photon pairs could be generated directly into a bandgap, whilst Raman shifted pump light becomes trapped in the bandgap in which the pump propagates due to the high loss region separating the pump from the idler. It still waits to be seen how much dispersion control can be exercised in these fibres,

7.2 Future Outlook 145

Fig. 7.1 A photograph of the two photon pair sources, side-by-side on the bench *top*

but if sufficient control can be made, these fibres could be a significant improvement over other fibre designs.

One key measurement that is missing from this discussion is the Hong-Ou-Mandel interference between heralded single photons from each source. This would provide a more accurate measure of the indistinguishability of the heralded photons from the multiplexed system; this aspect is still on going. A large benefit of the integrated fibre sources constructed here is the potential of replacing the current PCFs with a new design iteration, without having to alter any of the filtering or switching. One potential work package could be to replace the current fibre with birefringent PCFs for pair-generation such as those developed by McMillan et al. [16, 17]. These fibres have already been proven to yield high heralded count rates, and were designed to produce a factorable two-photon state.

7.2.2 Multiplexed Photon Pair Sources

Active multiplexing of photon pair sources is a top contender for producing a pseudo-deterministic source of single photons. In the light of the simulation results displayed in Chap. 3, optimising the collection efficiency of the photons from the source medium into the switch network, and minimising the loss of the switch network will be key. Additionally, it was seen that moving to PNR heralding detectors immediately brings significant performance enhancements. Development of high efficiency, fast PNR detectors in tandem with faster and lower loss optical switches will be key to fully realising the potential of multiplexed devices.

Poor collection efficiency can be mitigated by using more sources, but only if the switch loss is very low. Rather than the ideal 17 sources for a pseudo-deterministic source, tens of sources may be required if the loss is not improved (but this requires near perfect switches). The ability to fabricate large numbers of identical sources will be key. This is one of major the benefits of chip-scale waveguide sources. However, to minimise loss and remain an integrated device will necessitate developing switch, delay and detector technology in the host material, none of which are trivial to achieve. For the time being, fibre based sources have the potential to be scaled up to multiple sources with relative ease, especially if multiplexed in the temporal domain. For the immediate future of the sources fabricated in this thesis, moving into the temporal domain with the addition of a storage loop to the output is currently being explored.

An integrated source architecture offers many benefits over free-space implementations, primarily higher stability, portability and the potential for packaging into a commercial device. The large background in commercialisation of fibre based technology in the telecommunications industry will be highly advantageous. There are still some hurdles to overcome though. As was seen in Chap. 4, sources based on the current PCF design may be restricted by structural and dispersion fluctuations, but this could be addressed by redesigning the fibre to exploit birefringent PCF. The pitch and hole size required to achieve phasematched wavelengths of 810/1550 nm in such a scheme are much more routinely fabricated. As every length of fibre used in every source must be identical makes this difficult but not impossible to achieve, especially with the huge benefits that SET brings for characterisation.

Multiplexing in the temporal domain is well suited to fibre based sources, as the necessary storage loop can be easily fabricated in low loss single mode fibre. This also reduces the overheads in the number of identical PCF fibres required. As seen in the more recent literature [9], it is likely that some form of hybrid scheme, utilising both spatial and temporal multiplexing will prevail as a balance between performance and resource costs.

References

1. A.L. Migdall, D. Branning, S. Castelletto, Tailoring single-photon and multiphoton probabilities of a single-photon on-demand source. Phys. Rev. A **66**, 053805 (2002)
2. T.B. Pittman, B.C. Jacobs, J.D. Franson, Single photons on pseudodemand from stored parametric down-conversion. Phys. Rev. A **66**, 042303 (2002)
3. E. Jeffrey, N.A. Peters, P.G. Kwiat, Towards a periodic deterministic source of arbitrary single-photon states. New J. Phys. **6**, 100 (2004)
4. X.-S. Ma, S. Zotter, J. Kofler, T. Jennewein, A. Zeilinger, Experimental generation of single photons via active multiplexing. Phys. Rev. A **83**, 043814 (2011)
5. M.A. Broome, M.P. Almeida, A. Fedrizzi, A.G. White, Reducing multi-photon rates in pulsed down-conversion by temporal multiplexing. Opt. Express **19**, 22698–22708 (2011)
6. M.J. Collins, C. Xiong, I.H. Rey, T.D. Vo, J. He, S. Shahnia, C. Reardon, T.F. Krauss, M.J. Steel, A.S. Clark, B.J. Eggleton, Integrated spatial multiplexing of heralded single-photon sources. Nat Commun. **4** (2013)

7. T. Meany, L.A. Ngah, M.J. Collins, A.S. Clark, R.J. Williams, B.J. Eggleton, M.J. Steel, M.J. Withford, O. Alibart, S. Tanzilli, Hybrid photonic circuit for multiplexed heralded single photons. Laser Photon. Rev. **8**, L42–L46 (2014)
8. F. Kaneda, B.G. Christensen, J.J. Wong, H.S. Park, K.T. McCusker, P.G. Kwiat, Time-multiplexed heralded single-photon source. Optica **2**, 1010–1013 (2015)
9. G.J. Mendoza, R. Santagati, J. Munns, E. Hemsley, M. Piekarek, E. Martín-López, G.D. Marshall, D. Bonneau, M.G. Thompson, J.L. O'Brien, Active temporal and spatial multiplexing of photons. Optica **3**, 127–132 (2016)
10. C. Xiong, X. Zhang, Z. Liu, M.J. Collins, A. Mahendra, L.G. Helt, M.J. Steel, D.Y. Choi, C.J. Chae, P.H.W. Leong, B.J. Eggleton, Active temporal multiplexing of indistinguishable heralded single photons. Nat Commun. **7** (2016)
11. W.P. Grice, A.B. U'Ren, I.A. Walmsley, Eliminating frequency and space-time correlations in multiphoton states. Phys. Rev. A **64**, 063815 (2001)
12. K. Garay-Palmett, H.J. McGuinness, O. Cohen, J.S. Lundeen, R. Rangel-Rojo, A.B. U'ren, M.G. Raymer, C.J. McKinstrie, S. Radic, I.A. Walmsley, Photon pair-state preparation with tailored spectral properties by spontaneous four-wave mixing in photonic-crystal fiber. Opt. Express **15**, 14870–14886 (2007)
13. S.A. Cerqueira, F. Luan, C.M.B. Cordeiro, A.K. George, J.C. Knight, Hybrid photonic crystal fiber. Opt. Express **14**, 926–931 (2006)
14. L. Xiao, W. Jin, M. Demokan, Photonic crystal fibers confining light by both index-guiding and bandgap-guiding: hybrid PCFs. Opt. Express **15**, 15637–15647 (2007)
15. V. Pureur, A. Bétourné, G. Bouwmans, L. Bigot, A. Kudlinski, K. Delplace, A.L. Rouge, Y. Quiquempois, M. Douay, Overview on solid core photonic bandgap fibers. Fiber Integr. Optics **28**, 27–50 (2009)
16. A.R. McMillan, J. Fulconis, M. Halder, C. Xiong, J.G. Rarity, W.J. Wadsworth, Narrowband high-fidelity all-fibre source of heralded single photons at 1570 nm. Opt. Express **17**, 6156–6165 (2009)
17. A. McMillan, Development of an all-fibre source of heralded single photons, Ph.D. thesis, University of Bath (2011)

CPSIA information can be obtained
at www.ICGtesting.com
Printed in the USA
LVHW02*1449040318
568593LV00001B/95/P